John W. Macvey, an internationally renowned expert and writer on astronomy and astronautics, is a member of the American Astronomical Society of the Pacific and the British Interplanetary Society, and is a Fellow of the Royal Astronomical Society. The owner of his own observatory, Mr Macvey is the author of *Alone in the Universe* and *Journey to Alpha Centauri*. He lives in Ayrshire, Scotland.

John W Macvey

Whispers From Space

Paladin

Granada Publishing Limited
Published in 1976 by Paladin
Frogmore, St Albans, Herts AL2 2NF

First published in Great Britain by
Abelard-Schuman Ltd 1974
Copyright © John W. Macvey
Made and printed in Great Britain by
Richard Clay (The Chaucer Press) Ltd
Bungay, Suffolk
Set in Monotype Ehrhardt

This book is sold subject to the condition that it
shall not, by way of trade or otherwise, be lent,
re-sold, hired out or otherwise circulated
without the publisher's prior consent in any
form of binding or cover other than that in
which it is published and without a similar
condition including this condition being imposed
on the subsequent purchaser.
This book is published at a net price and is
supplied subject to the Publishers Association
Standard Conditions of Sale registered under the
Restrictive Trade Practices Act, 1956.

To Ellen

How unreasonable it would be to suppose that, besides the earth and sky which we can see, there are no other skies and no other earths.

Teng Mu (*circa* A.D. 1200)

Contents

Illustrations	11
Acknowledgements	13
Preface	15

Part 1

1	From Out the Darkness	21
2	An Idea and Its History	25
3	The Universe Around Us	43
4	The Star Worlds	77
5	Creatures of the Stars	103
6	Antenna Versus Starship	120
7	The Stars Have Voices	131

Part 2

8	Searching the Void	149
9	Which of the Host?	162
10	Project Ozma	170
11	Towards a Cosmic Tongue	184
12	Other Techniques	200
13	From What Far Star?	215
14	Communication Philosophy	247
15	The Road Ahead	259

Epilogue	271
Bibliography and Further Reading	275
Index	279

Illustrations

1. Star cloud in Sagittarius, source of first discovered radio emanations from space
2. Great Spiral Galaxy in Andromeda showing also satellite galaxies NGC 205 and 221
3. Cygnus 'A' radio source
4. Artist's conception of 600-foot radio-telescope, Sugar Grove, West Virginia
5. World's largest radio-telescope, Muenstereifel, Germany
6. Antenna 'mirror' base of world's largest radio-telescope, Muenstereifel, Germany
7. The Difference in transit times between interplanetary travel and interstellar travel
8. The principle of time-dilation interstellar travel
9. The principle of generation star travel

Acknowledgements

A WORK of this nature can hardly be compiled without considerable assistance from others. In particular I wish especially to acknowledge the help accorded me by Miss M. J. Logan who typed the script while coping nobly with frequent changes and additions to it on the part of the author. I would also thank Mr Patrick Moore, O.B.E., F.R.A.S., and Mr J. S. Glasby, B.Sc., F.R.A.S., for several very useful suggestions. To Mr Ronald Russell go my thanks for his three intriguing illustrations to Chapter 5.

I am also greatly indebted to the Royal Astronomical Society and the Lick Observatory, Mt Hamilton, California, for permission to use Plate 1 and to the Hale Observatories of Mts Wilson and Palomar, California, for permission to use Plate 2 and Plate 3. To the American Museum of Natural History, New York, and to the United States Navy go my thanks for consent to include Plate 4 and also to United Press International, New York, for Plate 5 and Plate 6. I must also express gratitude to the National Aeronautics and Space Administration, Washington, D.C., and to Imperial Chemical Industries Ltd, London, for enabling me to include a number of excellent and appropriate illustrations.

By no means least I acknowledge with deep gratitude the invaluable and much appreciated assistance accorded me within the United States by my very good friend and counsellor Mr M. K. Parkhurst of New York. It is appropriate also that due tribute be paid to the publishers for their continuing interest, support, and friendly guidance.

To all others who by virtue of their support and encouragement helped this book along go my most sincere and heartfelt thanks.

JOHN W. MACVEY

Saltcoats, Ayrshire
1973

Preface

HALF a century or so ago one of the most fascinating and perplexing questions confronting mankind was the possibility, indeed the probability, that intelligent life might exist elsewhere in the solar system. At that time serious suggestions had been made that some of our neighbouring worlds around the sun might contain civilizations possessing highly advanced technologies. Was it in fact possible, to paraphrase the words of H. G. Wells, that other worlds were watching the affairs of Earth?

The passage of years, coupled with great strides in the techniques of observational astronomy, brought nothing but disappointment to the proponents of the idea. There was after all, it seemed, no intelligent life within the solar system, other than that existing upon our own planet. By way of consolation came the admission that plant life of a very low order might exist on Mars and the much remoter possibility of primitive organisms existing on Venus or on the moon. It was a consolation that did little to console – a substitute that was no substitute. How far removed were these pitiful concepts from the enthralling ideas of the recent past: how poorly Martian lichen life compared with the race of master engineers and super-scientists so beloved of the science fiction writer.

So after all mankind was alone! From the heaving, awesome fires of the sun to the frigid, dark isolation of Pluto it seemed that only one world, our own, could contain intelligent living beings. Was man then just another cosmic accident, some irrelevant, inconsequential and unintentional by-product, a meaningless speck of impurity in the crucible of the universe?

Men looked out to those other suns, the stars, pondering in their minds the possibility that maybe *there* kindred living creatures might be found. The theories of the past had not entirely precluded the idea. By now, however, a new cosmology had been born, and with its coming the Copernican revolution had ground to a dismal halt. Anthropocentrism was in power, and its dogmatic doctrines stifled all argument. Very few, if indeed any, of the stars could have worlds around them. The solar system was unique; man was unique – and he was alone!

Fortunately the honeymoon period of anthropocentrism was relatively short. Cosmology was clearly seen to have taken a wrong course, and within a few decades it returned to the line of reasoning from which it should never have parted. Once more the universe could be seen for what in truth it really is – a mass of suns in whose awesome depths might lie a myriad of other worlds: a universe of light and darkness, of heat and cold, of matter and space, a universe of the living as well as the inert.

Yet life existing in this incredible remoteness represented an entirely different concept, one far removed from that of life on planets such as Mars lying on our cosmic doorstep. This was life so far removed from earth and sun as to be almost meaningless. Here was a celestial mirage, which existed and yet did not. The barriers of time and distance were too great. Some day, perhaps, in an incredibly distant era, mankind might leave the solar system and head out towards the stars. That, however, was the future; this was the present.

It became clear that total impasse had been reached. There was no other life in the solar system, no life on planets so conveniently close to our own. Life might exist among the stars, but the distances were so vast as to make the mind reel and sufficient almost to destroy sanity. Was mankind therefore, throughout the entire foreseeable future and beyond, to be isolated, to be barred from communication with beings on other worlds?

If it had not been for the advent of a new technology this isolation might well have persisted. Instead there has arisen the possibility of a most enthralling compromise. We cannot go out to meet these beings. Neither perhaps can they come to us. Instead, let us then endeavour to send radio signals to them. Let us try to intercept and to understand the possible messages alien civilizations may be beaming our way. The idea is not fantastic, not something borrowed from the exciting wonderland of science fiction. Already Frank Drake and his colleagues at Green Bank, West Virginia, have carried out the first tentative experiments along these lines. The vision, faith, and courage of these men should be loudly acclaimed for they are the first pioneers in one of the greatest projects ever conceived by man. It should be obvious that success, even very limited success, is unlikely to come either easily or early. Indeed many years may elapse before even the first hint of success is forthcoming. On that day we will have passed a milestone but it will be only the end of the beginning.

We must never be unmindful of the difficulties, never heedless of the magnitude of the task. But neither should we be afraid to display optimism and resolution. Pessimism and doubt are poor partners to take along with us into the closing decades of the twentieth century.

Above and surrounding us lie the eternal stars – bright islands in the vast, dark ocean of space. Between these islands and whispering across the star-strewn wastes of the heavens may be passing even now the messages that speak of galactic empires, of celestial dynasties, and of strange events long past.

We must therefore seek to ensure that our world is not excluded from this wonderful communications network. Let us, too, have the ears to hear, the tongue to speak, the mind to comprehend. Perhaps the words of Otto Struve, director of the United States National Radio Astronomy Observatory, offer the most appropriate inspiration: 'Unless we try we will never know.'

Part One

1 From Out the Darkness

THE land lies sleeping under the enveloping mantle of night. Bright stars gleam like jewels from out the velvet darkness of the moonless sky. Beyond these points of celestial beauty, in depths frightening in their sheer immensity, lie realms powdered in stellar glory. And beyond again – for always there must be a beyond – the Milky Way trails its tenuous gown of stardust across the heavens. Low in the far north the ethereal streamers of the aurora borealis silently flicker and flare.

Starkly against the splendour of the night sky rears the great metallic lattice of a giant radio telescope, the huge inverted dish which is its 'ear' pointing upward towards the heavens. In a small, low building one hundred yards or so from the mammoth instrument a small group of men and women watch with unconcealed wonder as a recording needle traces its purposeful red path. For years now this device has spoken only one language – the eternal jargon of the stars themselves, natural, irregular 'messages' emanating from distant, lonely suns, or whispers from galaxies so remote that the mind reels.

But now at last there is a change. From one small particular region in the sky has come suddenly a new 'voice', a different voice. At set intervals of seventeen minutes come five pulses of energy, three long, two short. They are weak and dreadfully attenuated, yet the message they bear is quite unmistakable, the implications staggering. This is no meaningless galactic chatter, no indiscriminate babble of thermally excited atoms in the far and awesome reaches of space. This is an intelligent message arriving from these very depths – a message sent on its terrible journey years ago by living, thinking beings of unknown form.

The instruments are rapidly adjusted in an endeavour to pinpoint more exactly the source of this strange, alien intelligence. The signals, it appears, emanate from a star in the sprawling constellation of Cetus, the Whale. Now by virtue of finer tuning and higher amplification the signals are stronger. But the message is the same – three long pulses and two short, with a gap of seventeen minutes before the signal is repeated.

The leader of the small group breaks away from his companions, moves over to a nearby window, and looks out thoughtfully, almost serenely, into the depths of the night sky. For perhaps a minute he is silent. It is a silence charged with deep emotion. Without turning or even averting his gaze from the heavens he finally speaks.

'My friends,' he says quietly, 'we stand in the presence of history. What we have so long suspected is true. Mankind is not unique. We are not, after all, alone in this universe!'

As the opening to a science fiction story the foregoing paragraphs might not appear particularly inappropriate. To suggest, however, that within the foreseeable future an event along these lines might actually take place may well give rise to feelings of considerable scepticism. Yet, perhaps surprisingly, this is hardly less than the truth.

That it is possible today to make such an assertion indicates that a really revolutionary change has overtaken our cosmological beliefs and concepts in recent times.

Forty years ago the idea that intelligent, cultivated races might occupy planets orbiting other stars would almost certainly have been dismissed, probably very peremptorily, as the sheerest fantasy. At that time what has come to be known as anthropocentrism held almost undisputed sway. In the epic words of Sir Arthur Eddington, the noted British cosmologist, 'Not one of the profusion of stars in their myriad clusters looks down on scenes comparable to those which are passing beneath the rays of the Sun.'

At that time it was firmly held by nearly all eminent astronomers and cosmologists that our sun was almost the only star in the entire galaxy of the Milky Way likely to possess a planetary retinue. That a view such as this should have been adhered to so widely seems especially odd today. We can recall how essentially reasonable theories relating to the creation of the solar system had been completely discarded in favour of the highly artificial concepts suggested by Sir James Jeans and Harold

Jeffries. Fundamental to the Jeans–Jeffries concepts was the belief that a passing star came so close to our sun as to draw from it a vast filament of glowing gas. From this fiery filament, it was argued, were born the planets, satellites, asteroids, and meteors that constitute that assemblage we know today as the solar system.

Now it is known that the distance separating star from star is so vast that the probability of one star making so close an approach to another is extremely slight, so slight in fact that it can virtually be ignored. Thus, according to this theory, what led to the formation of the solar system was unlikely to have been repeated at any other point in the galaxy. In other words there could be no possibility of other planetary systems and therefore no possibility of alien life, intelligent or otherwise.

Today the wheel of cosmological philosophy has gone full cycle and, having done so, may now have come permanently to rest. The older, sage and less spectacular ideas of Immanuel Kant and Pierre Simon de Laplace have been re-examined and re-appraised. After certain essential modifications (and a lot of improved mathematics!) we are left with the currently acceptable premise of C. von Weizsäcker. Here we have a theory that, while not entirely perfect, certainly does stand up reasonably well to rigorous testing. Other modifications to this theory may yet come about, may even be vitally necessary, but it now seems unlikely that in its general form it can or will suffer much change. In this concept, like those of Kant and Laplace to which it can be regarded as a direct descendant, the formation of a planetary system is not considered in any way as a unique event or as one peculiar to a single star known as the sun. On the contrary, the creation of planetary families can now be fairly and safely accepted as a natural phase in stellar evolution.

If, therefore, countless other planets do exist, it is hardly extrapolating our case too much to assume that at least a small proportion will be able to nurture and maintain advanced, intelligent life forms. The gulf between their worlds and our is immense by any standards. Some day we (or they!) may have the technology that will enable us to bridge this gulf physically. Clearly, interstellar travel, so far at least as we are concerned, must lie at a very remote and future point in history. Must this very real and truly awesome time and distance barrier separate us for ever from our neighbours in the galaxy? The answer must surely be negative. Must it even separate us in the more measurable and foreseeable future? We believe that this should be only in the strictly physical sense. Radio waves flash out with the speed of light itself (186,000 miles per second). Can we not then use these means to establish contact with alien races located on other planetary islands within the same great and enveloping galactic ocean?

Obviously the technical difficulties will be appreciable. Equally great must be the problems relating to language, outlook, and customs. But in time, with almost infinite patience and sincere endeavour, these obstacles could, we believe, be overcome.

These peoples, these strange other beings with whom we may eventually make electronic contact would, we must suppose, be at least our technological equals. It is not improbable that some, if not most, might even rank as our superiors. What such races might eventually teach us could prove of infinite and inestimable value – the remedies for disease and pestilence, the secret of star travel, even one day, perhaps, panaceas for the sicknesses of war and intolerance.

In the ensuing pages we will seek to examine as many relevant aspects as possible. We hope the tale will prove even half as fascinating as the subject with which it is concerned, for here surely is part of the shape of things to come.

2 An Idea and Its History

IT IS almost impossible to say accurately when man first began to consider the prospect of life on other worlds. To pinpoint the origin of this idea would make it necessary to examine the course of astronomy through the ages.

Astronomy had its roots among primitive peoples whose existence depended to a very large extent upon the dictates of nature. These peoples had neither clocks nor calendars but instead used the regularly changing face of the eternal heavens. The seasons for sowing, reaping, hunting, fishing, and so on were conveniently marked by the rising in the east of particular stars and constellations just as, for example, the rising of the well-known Pleiades indicates to us the approach of winter. There was among these people, of course, no understanding either of the motions of the various heavenly bodies or of their physical states. Indeed these bodies were often regarded with superstitious reverence and even at times with fear and awe. Thus was born a cult of mysticism that tended more towards astrology than astronomy. It was a cult that led not infrequently to horrifying and barbaric rituals. Yet from this paganism came some good, for its practice initiated systematic observation of the heavens.

The ancient civilizations of Babylon, Egypt, and China maintained, developed, and to some extent 'civilized' the practices the earlier peoples had begun. Observational methods and techniques were considerably improved, and from the accumulation of records it became possible to formulate certain basic rules that in turn led to computation and prediction on a wider and much more accurate scale.

However, it would be untrue to state that genuine astronomy had yet been born. Among the Egyptians astrolatry, whose fundamental purpose was to worship the stars, could be said to have preceded astronomy. Observations were made purely to facilitate the advancement of this worship, although they were not without some measure of astronomical value.

The Babylonians observed the heavens assiduously and were responsible for the division of the sky into the twelve well-known regions of the zodiac. In addition they were largely responsible for naming the constellations. They kept a close watch on the planets Mercury, Venus, Mars, Jupiter, and Saturn even though they had no idea or means of knowing that these were other worlds whose forms in some respects were akin to that of the planet Earth. Indeed at this stage in the history of man the very form of his own planet was a matter of considerable conjecture.

The process that led to devising the star figurations we know as constellations must have been one that developed slowly. It is now freely conceded that only a very few of these star groupings resemble the creatures or objects after which they were named. Allowance must, however, be made for the fact that many of the tribes in this part of the world were nomadic, living in the open permanently under skies almost perpetually clear. Above their tents and encampments blazed the whole panoply of stellar glory – a myriad of glittering points of light against the impalpable blackness of the sky. In that age, free from fouled atmospheres and the glare of artificial light, they would almost certainly have been able to see with the unaided eye many faint stars that could not be discerned in this fashion today. As these men gazed into the tranquil star-strewn depths it is perhaps not surprising to us that there they were able to see in all their glory many of the fabulous creatures and figures of legend.

The Babylonians appear to have been fairly adept in predicting eclipses of the sun and the moon. These people were also, it seems, aware of the fact that Venus returned to a given position in the sky after a period of eight years, and apparently succeeded too in predicting the future positions of Mercury, Mars, Jupiter, and Saturn.

The ancient Greeks continued where the peoples of the Near East left off, and it is probably from this quarter that there first arose the suggestion that the earth might be round and travel through space. Conceptions of the true nature of the solar system were, however, still incredibly vague. Heraclitus believed and taught that Mercury and Venus revolved round the sun which in turn orbited the planet Earth, a reasonable premise when one considers how Mercury and Venus appear near to the sun in terrestrial skies. Aristarchus, who might almost be said to have anticipated Copernicus, came closer to the truth by suggesting that *all* planets revolved round the sun.

Thereafter, astronomical thought, especially with regard to the possible existence of other worlds, was seriously retarded by the reactionary effects attributable to philosophy and theology, which were manifested over a considerable period of time – during the pre-Christian era, down through the Middle Ages, and indeed right up to the Renaissance.

Aristotle, despite his eminence as a philosopher, took a particularly reactionary stand by stating dogmatically that it was simply not possible there were other worlds in any way akin to earth. This may seem slightly less surprising when we recall the reaction of several eminent contemporaries only a short time ago to the subject of space travel! Aristotle offered little if any proof for his assertions, which virtually bordered on bigotry. There can be little doubt that in taking this stand he did a great disservice to astronomy because the ideas he adhered to and wrote of extensively continued to endure long after his death. Indeed, his ideas were resurrected almost a thousand years later by the Church to whose narrow outlook at that time they were especially suited. In fact the teachings and views of Aristotle achieved a pre-eminence then that completely surpassed the popularity they had known in the years immediately following his death.

In contrast, Plutarch, who seems to have cherished few illusions about the exalted place and status assigned to Earth, sought to establish our planet in its true and humbler role. He was also among the first to envisage the real nature of the moon. Admittedly his concepts were not entirely those of today, for in one of his books he peoples the moon with demons! Plutarch did, however, speak of lunar mountains, valleys, and plains.

A few decades after the death of Plutarch there appeared on the scene a book by Lucian of Samoas entitled *The True History*. The contents of this work certainly belied its title by being imaginative to a high degree. Indeed it is probably true to say that Lucian's volume ranks as the first example of science fiction.

The story concerns a ship and its crew that are lifted up by a gigantic whirlwind while cruising to the west of the Pillars of Hercules, better known as the Straits of Gibraltar. After seven dreadful days during which ship and crew are tempestuously hurled around the sky, the vessel is cast upon the surface of the moon – a means of transference only slightly more fantastic than a few of the weird schemes dreamed up by some science-fiction writers today! Lucian's moon is mercifully void of Plutarch's celebrated demons, containing instead a race of beings who, when not cavorting around on the backs of multi-headed birds, are preparing for war with rather improbable creatures whose home is the sun. For a long time afterwards no more works of this kind appeared. It is likely that had they done so both books and writers would have been consigned to the flames!

The era of the so-called Dark Ages certainly lived up to its name so far as astronomy was concerned. No real advances were possible while the views of Church and science remained so unreconciled. There was but one world – earth. Thoughts to the contrary were best left unexpressed – in the interests of the thinker's safety! Eventually, however, a reconciling formula was conceived by Thomas Aquinas. The impact of this reconciliation was quite astounding because the Church had performed an almost immediate *volte-face*, declaring that since the creative powers of God were limitless, other worlds must surely exist. There must have been a plentiful supply of strong stakes and dry faggots for those people unable to perform the about-turn in good time!

Until the closing years of the sixteenth century Aquinas's teaching and its application remained the accepted viewpoint. Then began the longest phase in the protracted conflict between religious dogma and scientific truth. In fairness it must be stated that what has been termed religious dogma was for the most part quite irreligious, being compounded largely of superstitious fear and political expediency.

Three well-known names appeared on the scene at this time: Copernicus, Galileo, and Kepler. The first, Nicolaus Copernicus, was born in 1473 at Thorn, in what was then Prussian Poland. He studied mathematics and astronomy at Cracow and law and medicine in Italy. In 1505, he was appointed physician to the Bishop of Ermland at Heilsberg in northern Germany. His dissatisfaction with Ptolemaic doctrines appears to have arisen early in his life. In 1517, he produced while at Heilsberg the first draft of his celebrated *De Revolutionibus Orbium Coelestium*, but it was not until 1543 and just prior to his death that the work was finally published. Copernicus's conception of the solar system differed radically from that of Ptolemy. He maintained, rightly as we all now know, that the sun was the central and principal body of the solar system, that the earth turned daily upon its axis, and that the stars, extremely remote, were virtually fixed so far as the earth's motions were concerned.

Copernicus was well aware that his theory was of a revolutionary nature and that it would almost certainly bring down the Church's wrath upon him. It was most likely for this reason that publication was delayed for over a quarter of a century. However, the anticipated onslaught did not develop until several years after the death of Copernicus. It would probably be wrong to attribute this respite to tolerance on the part of the Church hierarchy. Copernicus's theory was severely mathematical, and it is likely that for a period of time the full implications of the theory were not clearly understood in that quarter. It was Galileo, in fact, who reaped the whirlwind that Copernicus began.

The hypothesis put forward by Copernicus had the compelling attribute of simplicity, a factor by now sadly lacking in the Ptolemaic conception that was becoming increasingly more ponderous, complex, and utterly untenable. Moreover, Copernican theory constituted a practical basis upon which could be built a true representation of the entire universe.

Galileo Galilei was born at Pisa in 1564. At the early age of seventeen he entered the university there. At nineteen he discovered that the regular stroke of the pendulum could be utilized in the accurate measurement of time. In 1592, he was appointed professor of mathematics at Pisa, and while occupying this post he propounded his famous theory that all falling bodies, irrespective of size, descend at the same speed.

Galileo was responsible for the invention of many scientific instruments, and he designed in 1609 a new type of refracting telescope. This was a significant event in his life because it led him to make a series of astronomical observations in the course of which he discovered the existence of sun spots and the four major satellites of Jupiter (Io, Ganymede, Callisto, and Europa). His early inclinations had tended towards acceptance of Copernican theory, then regarded by the Church as heresy. The observations he had made confirmed him in the belief that Copernicus was right, and in 1613 he announced his support for the new theory. The Church reacted at once by warning him that if he continued it would be at his own peril. Nevertheless, in 1632 he published a book entitled *Dialogue on the Two Principal Systems of the World* in which once again he proclaimed his belief in Copernicus's concept. He was arrested, tried by the infamous Inquisition, and compelled under duress to recant. According to popular legend, however, he is reputed afterwards to have said quietly, *E pur si muove* ('Nevertheless it does move'), referring, of course, to the movement of the earth around the sun. For his alleged heresy he was condemned to imprisonment, but this was commuted to the milder punishment of house confinement in Florence. It was there that Galileo continued his researches up to the time of his death.

One cannot but be appalled at the blind bigotry of the Church whose fear of new ideas was so strong as to verge on psychotic obsession. Though it is easy to be wise in retrospect, the indisputable fact remains that theory of the Ptolemaic system had undergone so much modification that it is doubtful if by then even Ptolemy himself would have recognized it or, in view of Galileo's discoveries, could any longer have accepted it.

The degree of impediment offered to astronomical progress by the Church of that period is practically inestimable. It would be comforting to think that such bigotry was peculiar only to past eras in the affairs of men. The practice of brain-washing among certain contemporary totalitarian regimes unfortunately and clearly indicates that humanity does not always learn from past mistakes.

The Ptolemaic concept of the solar system was based on the idea that the earth occupied the central point around which the planets moved in a complex pattern of small circles, generally referred to as epicycles, the centres of which moved in larger circles. Aristarchus and Copernicus, however, regarded the sun as the central body around which the earth and the other planets moved in epicycles.

It was left to the third of the triumvirate, Johann Kepler, to establish the correct order of things. Kepler was born in 1571 in Württemberg. He became assistant to Tycho Brahe, the noted Danish astronomer, at Prague in 1600, after having been dismissed from a professorship at the Austrian University of Graz two years earlier because he allegedly had Protestant opinions. Brahe died in 1601, and Kepler then succeeded him as mathematician to Rudolf II, the Hapsburg emperor.

In the course of an outstanding career Tycho Brahe had carried out systematic and careful observations of the planet Mars. When the results of these were closely scrutinized by Kepler, he discovered that no matter how he arranged the major circle and epicycle it was impossible to reconcile observational evidence with existing theory. What did gradually become evident, however, was the fact that Mars moved not in a circle or epicycle but in an ellipse. Consequently it soon became apparent to Kepler that all the planets moved in elliptical orbits with the sun occupying one of the two focal points of each ellipse. He further established the fact that as a planet approaches the sun it moves more quickly. These discoveries formed the basis of Kepler's celebrated First and Second Laws, and his Third Law established the relationship between the distance of a planet from the sun and the time taken by that planet to complete one revolution around it.

Kepler published his three Laws in 1609 and 1616, thereby providing an essential basis for the subsequent work carried out by Isaac Newton. It is hardly an exaggeration to say that Kepler's Laws mark the starting-point of modern astronomy.

Unfortunately the findings of Copernicus, Galileo, and Kepler proved to be insufficient to overthrow established dogma. Although these findings certainly presaged the shape of things to come, they led also to an even more fanatical defence of the old order whose adherents became more deeply entrenched and embittered. We have already seen how when threatened with dire consequences Galileo was forced to make a retraction, albeit with considerable mental reservations. This was a mild fate compared to that reserved for an unfortunate Dominican monk named Giordano Bruno. Rather a boastful man, Bruno had accepted quite openly the views of Copernicus as he poured scorn on the Aristotelian view. For his sins he was imprisoned for seven years. Finally in 1600 the Church excommunicated him, and later he was burned alive at the stake. During his trial, if such a biased farce can be called a trial, Bruno spoke words that have since become immortal. 'Perhaps,' he said, 'you who condemn me are in greater fear than I who am condemned.' It is quite impossible not to wonder just how many supposed adherents of the old order had doubts in their hearts, doubts to which they dared not give utterance.

In the years that followed, it became increasingly clear that Copernicus, Kepler, and Galileo had not laboured and argued in vain. Nor perhaps had the courageous if foolhardy Bruno died in vain. Little by little the new ideas began to take hold. Progress at first was almost imperceptible. The process of conversion to the new order might be likened to the waves of the sea slowly undercutting a great cliff. For years little apparent inroad into the stubborn and seemingly inviolate rock is made. After a time, however, it becomes obvious that very small pieces are flaking off, and in due course it is clear that the whole face of the cliff is slowly being undermined. At last there comes a final great storm and, accompanied by a thunderous roar, the entire face of the cliff slides to destruction.

In this context the slow erosion caused by the little waves was apparent in several ways. Not the least of these was the appearance of a number of fictional works that played around with the possibility of life on other worlds and with the means of getting to them. These tales may seem childish and ludicrous to us, but it must be remembered that they bear the imprint of their own time and ought to be judged only on that basis.

In 1638, Bishop Godwin wrote a book in which the central figure of the story is towed to the moon in a sleigh-like contraption hauled by wild swans. We must remember that at this time it was believed that man had only to gain the power of flight to enable him to reach the moon, planets, and indeed the stars. The fact that the atmosphere of Earth was not a universal medium pervading all space was doubtless not even suspected.

The following year another bishop, this time an Englishman named Wilkins, produced a non-fictional work concerning the moon in which he raised the question of whether it was inhabited by intelligent beings. Here again we may be tempted to smile, but in fact we should not. Today we know the moon to be a stark and barren place. The very fact that the suggestion that the moon might be inhabited had been made at all showed that in some quarters at least it was suspected that the moon was a possible sister world of Earth, which it is within certain very narrow limits.

No reference to this era could be considered complete without reference to the almost legendary figure of Cyrano de Bergerac, who produced two early examples of what has come to be known as space fiction. In 1657 his *Voyage to the Moon* was published, to be followed five years later by *History of the States and Empires of the Sun*. Cyrano in the first instance gets round the difficulties inherent in space propulsion in a devious and dubious way. Dew on the ground soon evaporates or rises in the warm rays of the morning sun. Why, then, should not flasks of dew attached to a man or a vessel also rise, bearing their load with them? It seems doubtful if this technique would be greatly appreciated around the launching pads of Cape Kennedy, despite its economic attractions! Even the great Cyrano must later have had misgivings, for he goes on to dwell on the equally impracticable idea of an iron chariot somehow pulled skyward by pieces of lodestone. He also abandons this technique in favour of using gunpowder, a system that we would today describe as solid fuel propulsion.

The moon was an obvious first choice when it came to the question of alien life or extraterrestrial travel. It was left to Bernard de Fontenelle (1657–1737) to venture further afield. The work of this writer could almost be said to represent the culmination of what had gone before. Although his book, *Conversations About the Plurality of Worlds*, appeared in 1686, less than a century after the ceremonial incineration of the unfortunate Bruno, it was received with considerable acclaim and with not even the hint of a suggestion that its writer should become the centrepiece of a bonfire! In the years following the demise of Bruno, the erosion of the cliff of dogma had proceeded apace.

Though by present-day standards Fontenelle's knowledge of the planets is sketchy, this book must also be viewed in the light of its own time. Some of the author's planetary conceptions are not at all wide of the mark. For example, he envisaged Mercury as a very hot planet on whose surface ran rivers of melted gold and silver. Had he replaced these with pools of melted lead or tin he would hardly have been inaccurate. Jupiter he regarded as a very large planet, a judgement that we can in no way dispute, and he considered Saturn to be a very cold place. During this time the latter was the outermost known planet, which was probably the reason for this quite accurate belief. The moon was regarded by the writer as having an exceedingly thin atmosphere, and had Fontenelle merely extrapolated this idea further he would have been right yet again.

Fontenelle attributed people to each planet but showed surprising regard for ecological ethics by assuming that each people developed according to planetary environment. Admittedly, by placing intelligent beings on Mercury and on Saturn he went too far, but since this has been done repeatedly by fiction writers in the not too remote past every excuse can be made for Fontenelle.

Similar books appeared periodically between 1650 and 1800. Meanwhile the foundations of modern astronomy were being ever more firmly established. The invention of the telescope in 1608 had placed a hitherto unimaginable power in the hands of the astronomer. In 1668 the position was further enhanced by Sir Isaac Newton's invention of the reflecting telescope in which a concave mirror takes the place of the objective lens. This principle is found to this day in the great optical giants that rear their huge frames towards the heavens on the summits of Mt Palomar, Mt Wilson, and Mt Hamilton. Newton of course must be remembered most of all for his great theoretical work, *Principia*, published between 1686 and 1688. At last the mechanics of the solar system and the universe were beginning to be understood. Newton had explored the fringe waters of the great deeps that the incomparable Einstein would later plumb.

In 1781, entirely by chance, William Herschel discovered the formerly unknown planet Uranus, a great world lying beyond the lovely ringed Saturn. Soon irregularities in the motions of Uranus were pointing to the possible existence of yet another giant planet lying still farther out. The last remnants of the cliffs of dogma were crumbling fast.

By the early years of the nineteenth century, the possibility that mankind was not alone in the solar system was finding increasing support. A genuine appraisal of the prevailing conditions of the surfaces of the various planets had yet to be made, but something of the real nature of the lunar terrain was by then understood. It was thought, justifiably perhaps, that conditions on several of the other planets might permit the existence of advanced forms of life. There were of course then, as now, many who remained firmly convinced that only on Earth could life prevail, that the idea of reaching or communicating with other worlds was as impossible as it was ridiculous.

In 1865 there appeared Jules Verne's famous *Moon Voyage*, a book that makes fascinating reading to this very day. Here again is a work that should be read not with an eye to possible scientific inaccuracies (of which there are several) but in the light of its time. Just why Verne used a gigantic cannon to hurl his intrepid adventurers towards the moon is not clear, since it must have been apparent even to him that the human frame could not be expected to withstand such tremendous acceleration. That he understood the potential of the more orthodox rocket is shown by the fact that he makes use of braking rockets at a later point in the story. The use of the cannon as an adjunct to space exploration should not, however, be entirely ruled out. In 1964 Canadian scientists using an orthodox sixteen-inch rifled naval gun fired a projectile to a height of 100 miles. Clearly there is some scope for the use of a huge cannon as the first-stage carrier of an unmanned satellite or space probe.

As the nineteenth century drew to a close, increasing knowledge of the solar system began to render the idea of life on planets other than Mars and Venus more and more invalid. In the case of the two named planets, however, the possibilities were regarded as very strong. This was especially true in the case of Mars, the fabulous red planet whose very name in time became synonymous with a race of superior beings who might even one day descend on Earth. This belief had been fostered by a peculiar linguistic error. In 1877, the Italian astronomer Giovanni Schiaperelli first noticed what was apparently a series of straight lines traversing the surface of Mars, many of which seemed to run in pairs. Just what Schiaperelli himself thought has never been very clear, but in making the announcement of his discovery he used the Italian word *canali* ('channels'). This was most unfortunate because inevitably the word was translated into English as 'canals', thereby implying artificially created waterways. Canals covering the entire surface of a planet by then known to be seriously deficient in water could only mean the existence of a super-race of technologists intent on irrigating the planet by drawing water from the polar caps. Oddly enough a number of reputable astronomers hastened to support this concept. We even find Richard A. Proctor suggesting, in 1881, that on Mars there existed seas and oceans. In his *Poetry of Astronomy* he writes:

> Undoubtedly wide seas and oceans with many straits and bays and inland seas exist on Mars. We see in the telescope long white shore lines, the clearing mists of morning, the gathering mists of night – and we know that there must be air-currents in an atmosphere undergoing such changes. There must be rain and snow and hail, electrical storms, tornadoes and hurricanes.

Many astronomers, both amateur and professional, claim to have seen the so-called canals or at least markings of a canal-like nature. The late Professor Percival Lowell was so convinced, and his imagination so fired, that ultimately he devoted most of his life to the study of Mars. At Flagstaff, Arizona, an area renowned for the clarity of its atmosphere, he built an observatory, equipping it with one of the finest refractor telescopes of the day. It was here that Lowell produced his almost legendary map of Mars, a map criss-crossed with scores of straight lines. There was no doubt whatsoever in Lowell's mind that these lines represented vast canals produced by an incredibly gifted race. He was at all times a most meticulous observer, and there should be no doubt of his integrity. Other surface detail that Lowell recorded about Mars has stood the test of time and remains to this day as eloquent testimony to one of the finest and most gifted observational astronomers who ever lived. In the matter of the canals, however, did Lowell's immense enthusiasm and his tremendous desire to see them lead him into deluding himself? This may have had some bearing on the question, but since other astronomers, too, have caught a glimpse of these straight lines and on occasion even the camera has captured them, we must assume that they are not entirely illusory.

At the turn of the century, therefore, the concept of Mars as the abode of a superior intellect was finding wide acceptance. One or two authors were quick to make use of such an enthralling theme. Pride of place must go to the late H. G. Wells whose celebrated *War of the Worlds*, completed in 1897, probably still ranks as one of the greatest pieces of imaginative fiction ever written. At that time any phenomenon taking place on the surface of Mars was reported in considerable detail, and it is interesting to note that as a basis for his story Wells makes use of a report that actually appeared in *Nature*, the well-known scientific weekly journal. In the issue of 2 August 1894, there appears on page 319 a short item entitled 'A Strange Light on Mars', which describes flashes of light that occurred on Mars and were observed at the Nice Observatory by P. Perrotin. The article examines the possible reasons for the occurrence, and although it does not seriously suggest that the Martians were signalling, it does accept at least the possible existence of Martians, something that today no responsible account would do.

For the purposes of his story, Wells explained that the flashes were the discharges of a mighty cannon as it launched an Earth invasion force. The Martians when they reached Earth turned out to be loathsome creatures similar to octopuses. These creatures were at an immediate disadvantage in the oxygen-rich air and heavier gravity of Earth, but with their vastly superior technology they were soon able to overcome these difficulties. With their great tripod fighting machines and their terrifying heat ray, they were soon in almost complete control of Earth, whose cowed population was reduced to pitiful straits. All Earth's weapons were of little or no avail against the Martians. When this book was written, horse-drawn field artillery and the Maxim gun were the mainstays of national arsenals and so in the circumstances the result is perhaps hardly surprising. The only victories scored against the Martians' great machines were pyrrhic ones gained by naval vessels armed with huge rifled guns. (It is interesting to record that when a rather excellent film version of the book was produced in the United States a number of years ago the producers gave the story a contemporary setting in which tanks and jet bombers were employed against the Martians. However, even the power of the Strategic Air Command and the thermonuclear bomb proved insufficient to check the invading force.) At the close of the book (as in the film), the Martians are finally slain by an unspectacular but deadly adversary against which they have no protection and upon which they had not counted – the legion of invisible microbes to which over thousands of years terrestrial man had become immune.

In 1901 Wells produced another of his science fiction classics, entitled *The First Men in the Moon*. This book in a sense is less straightforward than *The War of the Worlds* because it contains a moral, or one might even say an indictment of our civilization. In this fictional work the moon is inhabited by a race of creatures known as Selenites. It is unlikely that Wells, himself a science graduate, really believed the moon to be inhabited or indeed habitable. The moon, however, served as a convenient locale in which Wells could illustrate the point he had in mind.

He wrote about a method of reaching the moon that was simple but improbable. Cavor, an inventor, produced a material capable of nullifying the influence of gravity. When this material was incorporated in the space vessel, Cavor and his friend Bedford were able to rise effortlessly from the surface of the earth and steer a course for the moon.

The Selenites live *within* the moon and are a race of giant intelligent insects. Gold, being a common metal on the moon in the fiction of Wells, is used to much the same extent that we use the ordinary metals of Earth. Not surprisingly Cavor and his associate are impressed by this state of affairs. The Selenites, a cultured and peaceful race despite their form, are appalled by the attitudes and customs of early twentieth-century Earth. (It is interesting to reflect on what their reactions might have been to *late* twentieth-century Earth!)

The Selenites become increasingly fearful of the violence and avarice inherent in the nature of their nearest cosmic neighbours, and eventually Cavor and his friend find themselves in peril. The latter escapes and returns to Earth, where the space sphere is lost along with the secret of the fabulous antigravity material. Cavor remains on the moon as a prisoner of the Selenites. For a time he is able to send messages to Earth, but these eventually cease.

Venus, also considered for some time as possibly being inhabited by living creatures, never attained the popularity of Mars with either fiction writers or the man in the street. Although nearer to the Earth than Mars, Venus remained an enigma because of its permanently cloud-enshrouded surface. The face of Mars could be seen and charted even though its peculiar markings could not be adequately explained. There were deserts or great plains, probably high ground or plateaux, and certainly white polar caps that waxed and waned with the changing Martian seasons. There were also clouds of a kind, there was evidence of sand storms, and last but not least there was the possibility of canals. Every two years Mars could be easily seen against the dark night sky as a bright red 'star'. Mars was the planet of mystery, the probable abode of a fantastically advanced race. Men looked up and were pleasantly frightened. There above them was science fiction come true. Fact, legend, fancy, and speculation combined, and the word they spelled was *Mars*. Its very name had a ring that conjured up strange visions.

With Venus it was different. Astronomers told of the great cloudy atmosphere that consisted largely of suffocating carbon dioxide, but they admitted that the surface was permanently concealed from prying eyes. What lay beneath these clouds might be a rock-strewn desert, an all-embracing ocean, or a vast primeval swamp. The cosmology of the day suggested that Venus might be a younger world than the Earth and that there might be life in an early stage of development. Venus shone in the sky with a brilliance exceeding that of Mars and generally set quite soon after the sun or rose shortly before it in the morning. It never hung overhead near the zenith against a completely black sky. The fiction writers largely left it alone. No flashes or other intriguing phenomena had been seen to occur on Venus. All in all the lure of Mars was too strong. And so Venus, the lovely evening or morning star, failed to make that impression on the minds of men that the red world had done.

As the twentieth century progressed, the whole concept of the existence of life elsewhere in the solar system began to change. More powerful telescopes and fresh observational techniques revealed more and more of the secrets about the planets. With each disclosure these sister worlds of Earth became more stark, more forbidding, and less likely as habitats of life, intelligent or otherwise. Science-fiction writers continued to fill pages with adventures among Martians, Venusians, Jovians, Saturnians, and so on. Eventually, however, the truth had to be faced, even if it were unpalatable. Life demanded a critical and narrow band of physical conditions for its initiation and development. Only Earth satisfied these conditions. The rest were either too hot or too cold, too big or too small, had poisonous atmospheres or none at all. Was it ecologically possible for life to adapt itself to such conditions? At best, this was seen as unlikely. Man was alone, quite alone in the solar system!

Was he, however, alone in the universe? This was another question entirely, and one that at once raised many interesting and intriguing possibilities. By now it had long been known that the stars were merely other suns fantastically remote from our own. They were of various types and of varying ages. Many, however, were akin to our own sun and of similar age. Could planets exist around some of these stars? More specifically, could these planets contain, as does Earth, advanced civilizations? The main question that had to be answered in this context revolved around the evolution of planetary systems. Was the Sun's planetary retinue merely typical or was it unique? The remoteness of the stars precluded any possibility of detecting such planets optically, regardless of the size of the telescope employed. The question therefore could only be examined from a theoretical angle. Cosmologists for a time believed that planetary systems were probably formed as a matter of course, but eventually more radical beliefs began to hold sway. These beliefs by their nature implied that the birth of the solar system must have been an almost unique event. In such circumstances a widespread plurality of worlds became a totally invalid idea.

As it turned out, these concepts ran into increasing difficulties, which in the end became so overwhelming that they could no longer be ignored. Meanwhile a new understanding of stellar processes and life cycles had been steadily developing, and eventually the existence of innumerable planetary systems scattered throughout the universe was again accepted. A fuller treatment of this aspect will be found in Chapter 4.

By indirect means it was fairly and conclusively established that one or two stars were accompanied by large, non-stellar bodies (i.e., giant planets). Although large massive planets represent most improbable locales for life, the fact that they existed at all was a cardinal point. Where large planets exist so also may smaller ones, just as we find within the solar system the small planets of Venus, Earth, and Mars accompanying the giants Jupiter, Saturn, Uranus, and Neptune.

For some time science-fiction writers had been demonstrating an increasing tendency to move the scenes of their stories from Earth's lifeless sister worlds to hypothetical but possible cousin worlds orbiting other stars. The problem of reaching these distant worlds, as we shall see in Chapter 6, is truly stupendous, but such writers, never lacking in ingenuity, refused to let this be a great hindrance. Indeed the romance of traversing vast interstellar distances became one of the most exciting and fascinating features of such stories.

At this point it is interesting to note that one of the very earliest of such writers had already made a move in this direction. The writer was the renowned Voltaire (1694-1778), who published in 1752 a book entitled *Micromegas*, which featured the exploits of a giant from the star Sirius.

Today the solar system, except for Earth, is regarded as being void of life. In this context life must be interpreted as meaning intelligent and advanced living creatures. It is possible, indeed it is extremely likely, that a low order of plant life exists on Mars, perhaps in the nature of lichen or some moss-like growth. Also, we cannot rule out the likelihood, perhaps a little more remote, of some kind of insect life. Venus may contain some very early form of life, but this remains speculative in view of the high temperatures now known to exist there.

The idea of life with regard to the planetary systems of other stars is full of enthralling possibilities. Inevitably there must be much speculation about the forms such alien life might assume, but there now seems little doubt that after decades of conjecture, it can finally be said that man is *not* alone in the universe. It is a position of course that is unsatisfactory, indeed we might even say frustrating, because of the frightening time and distance barrier that separates us from our cosmic neighbours. It is not impossible that very advanced civilizations, having perfected the techniques of interstellar travel, may one day reach us. On the other hand, space is so incredibly vast and awesome that the chance that another people might select our sun as a target star is perhaps very remote. Moreover, would we, the peoples of Earth, really welcome representatives of a race that could very well be grotesquely different from our own?

If electronic contact can be secured with some of these alien races the outlook in this respect might be considerably changed. We could thereby learn something of the hopes, fears, and aspirations of other cosmic peoples and consequently enable the first rung in the ladder of amity and understanding to be attained.

Mankind has come a long way in his attitudes to other worlds – from the day he first pondered fearfully the possibilities of their existence to an era in which he prepares to set foot on the planets and reach out electronically towards the stars.

3 The Universe Around Us

In view of this book's theme it might appear more relevant to begin with the stars and their probable planetary retinues. Nevertheless we are going to start at a point much further out in both space and time. By so doing we may be able to present a clearer picture of the universe through which even now intriguing electronic messages may be passing.

Life is spawned by planets that in turn are born from stars. Stars in their turn owe their existence to great glowing clouds of hydrogen from which they are fashioned. But even these clouds of hydrogen do not represent true genesis. We must go still further back and outwards. We come to the galaxies, vast 'island-universes', shoals of stars, of heat, and of light in an endless ocean of cold and darkness. Our cosmic footsteps must in fact lead us to the very dawn of all history, to the moment of creation itself. Only in this way can we see the whole master plan, view the great sweep of God's inimitable artistry in all its fantastic and glorious immensity.

Clearly at some undetermined time and in some inexplicable way the universe must have come into existence. Just how (as well as when) is a question that has long fascinated cosmologists and philosophers alike.

Two theories more or less dominate this fundamental cosmological issue. The first of these is the hypothesis of the 'big bang' – a convulsive, primal cataclysm that supposedly brought the universe violently and suddenly into existence 20,000 million years ago. This theory was originally postulated by the Belgian mathematician, Abbé Lemaître. It envisions an incredibly distant epoch when the entire universe was jammed together in a sort of vast primeval 'atom', the diameter of which may have been in the region of 100 million miles. Within this titanic cosmic 'bomb' was packed all the matter from which today's universe has grown. At some incredible moment, for reasons we can never hope to know, the bomb went off. At that instant, which may have occurred some 20,000 million years ago, what we conceive of as time and space began.

It is futile to seek words that might even begin to describe the convulsive, cataclysmic fury of such an awesome detonation. If it happened, then it was simply a thing of inconceivable dimensions. The contents of this elemental cosmic thunderbolt were hurled violently outwards, and we are probably still witnessing the stupendous aftermath.

Let us use our imagination in endeavouring to follow this strange process through its ensuing stages. Four minutes after detonation the expanding fireball has attained the almost fantastic temperature of 1,000 million degrees and a radius of 39 light years. Two minutes later the temperature has dropped to 100 million degrees and the radius of the expanding sphere has become 390 light years – a tenfold dimensional increase. A month later the temperature has fallen to a mere million degrees, the sphere by now having a radius of several thousand light years. Thirty million years must elapse before we look at our explosion again. By now the temperature has dropped to a few thousand degrees and already a certain amount of gas has condensed into dust. Later still, gravitational effects result in the creation of chance accumulations of this dust that in due course become the gigantic clouds from which individual galaxies will ultimately be formed. A point strongly in favour of this theory, of course, is the fact that today the most distant galaxies are moving outwards with the greatest velocity. This would certainly seem to agree with the idea of an initial explosive outburst.

Today this idea of an explosive outburst is seen in many quarters as only part of the story. It may well be asked how the original cosmic bomb came to exist and whether the universe will continue to expand indefinitely. If we adhere to the idea of the single explosion these are difficult, if not impossible, questions to answer. If, on the other hand, we agree with the idea of an oscillating or pulsating universe, with the concept of a universe that contracts to a very dense, hot state before expanding again, such questions do not arise. It will be a good plan, therefore, to review our theory of creation in the light of this idea.

The strength of gravitational force is directly proportional to the masses on which it operates. So great, then, is the amount of matter in the galaxies that this gravitational force is a most important factor. Indeed, gravity could be said to dominate the motion of the universe. We have already mentioned that gravitational effects were primarily responsible for breaking up the expanding mass into the clouds from which the galaxies in time evolved. It also governs the motions of these galaxies and is chiefly responsible for holding them together.

If gravitational interaction were weaker than it is, the universe would probably continue to expand indefinitely. The amount of matter in space, however, leads to a gravitational value that

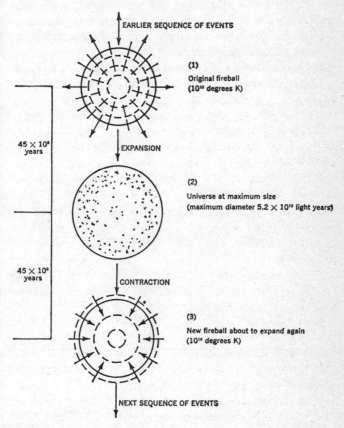

Figure 1a

should prevent this expansion from happening. Thus the universe will stop expanding when it has attained a certain critical size. It will thereafter contract to reconstitute a fireball of matter and radiation having a fantastically high temperature (probably in excess of 10,000 million degrees) and dimensions of the order of a few light years.

Why at this stage there should be such high temperatures may at first seem puzzling. There does exist, however, a possible and very simple explanation. Later in this chapter we shall see how the hydrogen within stars is gradually converted to helium and then in turn to heavier elements. In its early epochs our galaxy is known to have been free of such heavy elements. The implication, therefore, is that during the previous contraction the temperature rose sufficiently to reconvert these heavy elements to pure hydrogen, a process known to require temperatures in excess of 10,000 million degrees.

The probable cycle of events is shown in Figure 1a and the accompanying graph in Figure 1b gives an idea of the time scale involved.

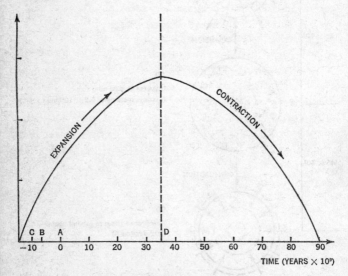

Figure 1b   A = present time
B = formation of solar system
C = formation of galaxy
D = maximum diameter (5.2×10^{10}: light years)

It was suggested by Dicke of Princeton University and others that it should still be possible to detect the radiation emitted by the fireball responsible for the present cycle of expansion. This, he postulated, should show itself as a uniformly distributed cosmic background at a wavelength of 3·2 centimetres. During 1964 Dicke and his colleagues designed a radio-telescope to function at 3·2 centimetres so that it could be ascertained whether or not such radiation could be detected. Their efforts were so successful that radiation was indeed found to exist. This constituted a truly remarkable discovery having tremendous implications. Radiation was reaching us from quite literally the dawn of creation, for these were waves generated in the great fireball only 200 seconds or so after the cataclysmic detonation!

What, then, of the ultimate future if our theory of a pulsating universe has validity? It is believed by some cosmologists that the universe will attain its maximum size of 26,000 million light years some 45,000 million years after the initial outburst. It will then begin to contract and eventually collapse catastrophically about 89,000 million years from now – a point in time of no possible concern to ourselves!

The second theory that seeks to account for the existence of the universe is the famous 'steady state' theory attributed to Hoyle, Bondi, and Gold. This theory contains the assumption that matter is continually being created as the universe expands so that the average density of the latter remains constant. The protagonists of this theory make no attempt to refute the undeniable evidence of a violent, outward motion on the part of the universe. They do, however, interpret this in a different way, substituting a form of repulsive force in place of Lemaître's explosion theory. The steady state theory is for the most part untenable today. The idea of matter creating itself out of nothing was always difficult to accept since it seemed contrary to all the dictates of reason. Moreover, fresh discoveries concerning the nature and dimensions of the universe have cast grave doubts even on its partial validity.

But now we must come forward in time – to the day when all the separate galaxies that constitute this vast universe exist. We come in fact to the present. Before we specifically examine our own galaxy, however, it is both sensible and expedient to take a quick searching look at the heavens in their entirety. Our galaxy, the Milky Way, is but one in a universe in which the number of such galaxies seems almost infinite.

When we survey the star-powdered immensity of the heavens on a clear and moonless night we are looking at one of nature's grandest and proudest sights. We see many bright stars, even more less bright, and some that are almost indiscernible to the naked eye. A very slight optical aid, such as a pair of good binoculars or a small telescope, reveals an even greater profusion of these points of light. An astronomical telescope of reasonable aperture greatly augments the viewing of this profusion. Soon we realize that every increase in optical power brings more of these other suns to our vision. Now, vast though the distances separating these stars from our own world are, on the great cosmic scale they are insignificant. Those stars we see with the naked eye and with minor optical aid belong for the most part to a local cluster or colony of stars within our own galaxy. Our sun, the fount of our very existence, is merely one of these stars. This assemblage of stars certainly does not represent the entire galaxy of the Milky Way. On the contrary, it is but a minute fragment of this cosmic colossus.

Let us return briefly to our dark, starry night. Sweeping across the great inverted bowl of the heavens is a faint misty band of light. When very powerful telescopes are directed towards this and long-exposure photographs are made, the plates reveal the Milky Way as a veritable mass of stars. This diaphanous band of light represents the major portion of our great galaxy. Our local group of stars in fact is merely one of a legion constituting this galaxy.

The reason why the main mass of the Milky Way should appear as a mere ribbon of diffused light trailing across the night sky is due wholly to its shape. It is in fact a vast, slowly rotating disc of stars, dust, and gas, the diameter of which is some 100,000 light years. Were we to see it from above (i.e., in plan view), it would display a spiral form strongly reminiscent of a very large and complex pinwheel firework. Seen from the side, the impression would be that of a tremendous lens suspended in space.

A few decades ago the structure of the Milky Way was to a certain extent a matter of supposition and speculation. Since then, however, radio-astronomers have been able to substantiate and extend much of the earlier reasoning. The interstellar gas located in the spiral arms of the Milky Way is a source of fairly strong radio-emission, and this has rendered it possible to draw up a much more accurate picture of our own galaxy. In the past it had been possible only to map the merest fragments of the spiral arms. Now, however, considerable portions of these arms can be accurately charted. Nevertheless, the picture presented today agrees in a large measure with the one predicted before the advent of radio-astronomical techniques.

An idea of the possible structure of our galaxy had been derived earlier from observations carried out on certain other galaxies. Astronomers had been able, thanks to the resolution of giant telescopes and the power of the photographic plate, to examine whole island universes in their entirety, something which, of course, it is impossible to do with our own. Amid the maze of nearby stars, clouds of obscuring dust, and swirling banks of elemental gas, we get at best a most imperfect impression of the great stellar metropolis to which we belong. In time and with considerable diligence, we have nevertheless been able to draw up a kind of chart. Although essentially accurate, it would be surprising if this chart did not contain some discrepancies. How convenient it would be to retire out into the great deeps of space and from there survey the Milky Way in all its splendour and infinite majesty. This, regrettably, we cannot do, but we are able to do something that, though less dramatic, is almost as useful. Unable to gaze upon our own galaxy from a point beyond its confines, we are already in that fortunate position with regard to other galaxies. These we can see and study in their entirety.

Since we are aware that many galaxies are akin to our own in structure and general configuration, we know that to study them is virtually to study the Milky Way. In effect these galaxies represent a sort of mirror in space. When we pause to reflect on the extent to which life probably proliferates amid the diffused star dust of the various island universes, it is not hard to envisage astronomers and cosmologists of remote worlds in these other galaxies surveying our Milky Way through their telescopes and seeing in it a facsimile of their own galaxy. In his renowned *War of the Worlds*, H. G. Wells posed the question, 'Are other worlds watching us?' To an extent that perhaps even this great visionary did not then realize, the answer may be 'Yes!'

Therefore we can look beyond the nearby star profusion of our own stellar system, out into the great black deeps of intergalactic space, and see there those vast stellar citadels that are the other galaxies glowing in their remote splendour. We have entered into a new and enthralling dimension, for now we see not another moon or another sun – we behold instead a universe!

If on a moonless autumn night when the sky is clear and the air free from mist and fog we look closely at the constellation Andromeda, we can just discern at one point a small faint patch of light. It looks rather different from the rest of the stars because of its somewhat hazy appearance. A swift glance at the constellation would almost certainly fail to reveal its presence. Nevertheless, this small, trivial-looking patch of light exceeds both in dimension and in importance all the other bright stars we see, for *this* is no mere other sun. It is a complete universe of perhaps 200,000 million stars, which is even larger than our own great Milky Way. We are looking at the Great Nebula in Andromeda, which, though $2\frac{1}{4}$ million light years remote, is one of our nearest neighbour universes. Here is an object so faint as to be almost invisible to the naked eye, an object sending light to us over the fearful abyss of 10 trillion miles. Here is light that started on its terrible journey before the dawn of our race, when the great reptiles were still lords of the Earth. For all that time, throughout all recorded history, it has been rushing towards us with a velocity too tremendous for our minds to conceive. With every passing second that light has spanned another 186,000 miles – and it has been doing that for every second contained in that $2\frac{1}{4}$ million years. Yet this is the *nearest* other universe to our own! It is a billion miles more remote than the moon, many thousands of millions of times more distant than the farthest planet of the solar system, half a million times farther from us than the nearest star, a thousand times farther than the faintest star that our unaided eye can discern. These are distances that make the mind reel, yet they are only the fringe waters of astronomy's greatest deeps.

This concept and acceptance of other separate universes lying far out in space is to all intents and purposes a fairly recent one.

Although some idea of the true state of affairs had been gaining ground for some time, it was not until 1925 that the renowned American cosmologist Edwin Hubble could positively announce to the world that other universes did in fact exist. During the preceding year, Hubble used the brand-new 100-inch Hooker reflector telescope of the Mt Wilson Observatory to detect distinctly separate stars in what for a long time had been regarded as mere clouds of gas *within* the Milky Way. One of these 'clouds' was our old friend the Great Nebula in Andromeda, or M31 as it is more generally known to astronomers.

It is doubtful if there will ever be an astronomical discovery of greater significance. Here at one blow the horizons of mankind were swept back to a fantastic extent. If the cosmic reaches had once been vast, now they were almost limitless. The bounds of space had opened wide.

It should perhaps be stressed at this point that not all the host of other universes are completely alike. Earlier we mentioned a similarity. This is no contradiction. All have a very great deal in common, and many are closely akin to the Milky Way and M31 in Andromeda. There are, however, distinct differences in size, shape, and orientation. M31, for example, is somewhat larger than the Milky Way but almost identical in most other respects.

The shapes that galaxies assume are quite varied. These range from amorphous, featureless bright clouds to magnificently symmetrical spirals. Moreover their orientation, the angle at which we see them, can also differ very considerably. This, of course, is not an inherent difference in the galaxy itself.

At first there was an element of confusion when it came to classifying galaxies. Hubble was eventually able to clarify the confusion by grouping them according to their respective shapes. Three distinct types of galaxy thus came to be known.

1. Spirals, a category accounting for some 80 per cent of the known galaxies;
2. Ellipticals, accounting for another 17 per cent;
3. Irregulars.

Spirals, the first group, are reminiscent of vast Catherine Wheels and are truly magnificent sights. The impression they give is one of tremendous majesty, of immense, sweeping, aeon-consuming motion. It is to this group that our own Milky Way and M31 in Andromeda belong. The distinctive features of this type of galaxy are the central nucleus or hub (H), a disc composed of the spiral arms (D) surrounding the hub in the equatorial plane, and finally a halo or 'shell' of star clusters (S) and stars surrounding the entire galaxy in the fashion of a sphere. (See Figure 2.)

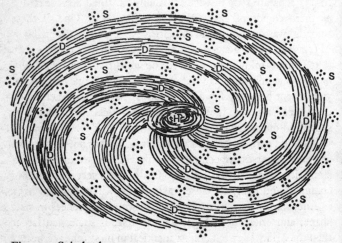

Figure 2. Spiral galaxy.

Elliptical galaxies have a rather less elaborate structure, in which there are really only two distinct portions – the hub and the spherical 'halo' of stars. (See Figure 3.)

Ellipticals are generally fairly large galaxies that are entirely composed of older stars, notably large red giants, variable stars, and dense white dwarfs.

Irregular galaxies, as the name implies, have ill-defined configurations in which the vague hint of spiral structure can sometimes be observed. (See Figure 4.) These galaxies are composed of intensely hot, blue and bluish-white new stars as well as vast amounts of dust and gas lying between the stars.

It might justifiably be thought that a single island universe should be about the largest unit we ought to find on the cosmic scene. However, we know today that this is not entirely true for it has been found that galaxies do in fact tend to group together into colonies of universes. In view of the terrifying and almost

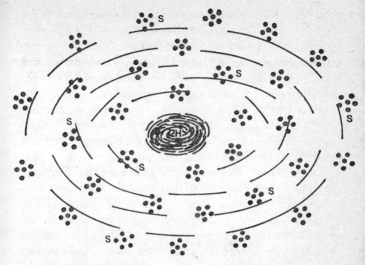

Figure 3. Elliptical galaxy.

unimaginable distances between individual galaxies, this might at first seem to be a totally absurd idea. Nevertheless, island universes do group together to form cosmic archipelagoes. We must at all times remember that in astronomical matters distance like dimension tends to be very relative. By strictly terrestrial

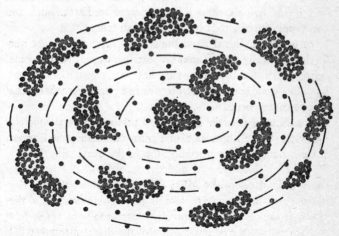

Figure 4. Irregular galaxy.

standards our moon is remote despite its dominant position in our skies. The distance of 240,000-odd miles contrasts very strongly with the 3,000 or so miles separating Europe from the United States. By similar reasoning, the moon is close compared to the planet Mars, which is still some 35 million miles remote, even when it is nearest to Earth. Pluto, the outermost known planet of the solar system, lying on the very fringes of interstellar space 3,000 million miles from Earth, is very remote indeed compared to the plainly visible Mars. Yet Pluto is close compared to the nearer of the stars, the latter being really quite near when we reflect on the vast dimensions of the Milky Way. It is hardly surprising, then, that all the stars in the Milky Way seem to be virtual neighbours when we think of the yawning gulf of $2\frac{1}{4}$ million light years that separates us from the galaxy Andromeda. But still this escalating process of comparison must go on. The Milky Way, M31, and several other island universes are merely constituents of a local galactic archipelago in the limitless, black ocean of space. At distances that defy all imagination we find on the far celestial horizon more of these strange archipelagoes. The most remote such group detected by the magnificent 200-inch Hale telescope on the summit of Mt Palomar lies in the constellation Coma Berenices. Its distance from us is 1,000 million light years! To a mountain top in California comes light that began its journey 1,000 million years ago, before living creatures had appeared on Earth. Such are the simple incontrovertible facts of astronomy that we must learn to accept. But if these dimensions appear to reduce our world to pitifully insignificant proportions, we may restore some sense of balance by considering how enormous Earth would seem to a small speck of microbic life in a pond, were that speck able to comprehend.

The facts of galactic distance are perhaps more adequately imparted by the illustration in Figure 5a. In the small, inner circle we find our own galaxy, the Milky Way, with the two Magellanic Clouds shown directly beneath it. The latter are really appendages of the Milky Way, or satellite galaxies. The radius of the inner circle represents 300,000 light years, and the radius of the outer circle represents 1,700,000 light years. Just beyond the outer circle at the top of the diagram (marked X) we find M31 in Andromeda and a few other attendant galaxies.

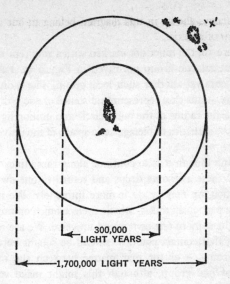

Figure 5a. Distance relation between Milky Way and nearest other galaxies.

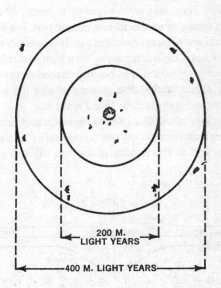

Figure 5b. Distance relation between local group of galaxies and nearest other groups.

55

All the galaxies shown in this diagram belong to our so-called local group of galaxies.

In Figure 5b, the inner dot marked with a ring represents the entire local galactic group portrayed in Figure 5a. Each other dot represents yet another such local group. The radius of the inner circle in this case represents a distance of 200 million light years, with the radius of the outer circle 400 million light years. Such are the statistics of metagalactic space, of the universe, and of infinity!

Tempting though it is to continue along these lines we must now leave these awesome deeps and return to our own galaxy in order that we may study it more intimately. We must not forget our particular thesis, which is electronic communication between the stars of our own island universe. We are not going to suggest that contact could or should be sought between the galactic groups – or for that matter between the individual galaxies of one group, although this might make very good science fiction. It is appropriate, however, that before discussing our main theme we should at least have some idea of the immensity surrounding us.

The Milky Way, that great wheeling universe, of which our sun and its family of worlds form a minute part, is a huge spiral galaxy composed of stars, dust, and gas. It is in effect a spherical system that has a central region of very high stellar concentration. This central portion or hub is surrounded on the equatorial plane by a rather flattish disc. Around all this lies a sphere or halo of stars and star clusters. (See Figure 6a.)

The diameter of the disc (AB) is approximately 100,000 light years, its greatest breadth, the central region (EF) is some 20,000 light years, and in the vicinity of our sun (S) it is 2,500 light

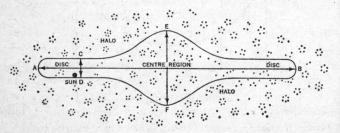

Figure 6a. Milky Way.

years (CD). Estimates of the number of stars contained in the galaxy vary considerably, ranging from 30,000 million to 100,000 million. The arms emerging from the central nucleus consist of what are termed Population I stars (of which the sun is a fairly typical example), blue giants and loose, open star clusters, and dust, and gas. The halo and hub are comprised of Population II stars, which are large red giants of gigantic dimensions and relatively low temperatures. Population II stars may also be found in the vacant spaces between the spiral arms. The halo also contains globular star clusters but no dust.

This whole system is in majestic revolution, but obviously speed of rotation so far as the individual stars are concerned must vary enormously depending as it does on the stars' respective positions within the galaxy. The stars at the exact centre of the hub have little or no radial motion, whereas those at the extremities of the hub have a fairly slow rate. On the other hand, those at or near the perimeter possess a very high degree of radial motion. The sun lies some 30,000 light years from the centre of the hub and 20,000 light years from the rim so that obviously its radial velocity will be considerable. An estimate has put this velocity at 135 miles per second. Consequently it must take about 225 million years to complete just *one* circuit. The last occasion, therefore, when our sun lay in its present position was during that remote period, long before the advent of earliest man, when the coal deposits were being laid down on our planet. Another 225 million years will bring it back to the position it occupies at the present time. Thus a historian in the remote year A.D. 225,001,976 will be able to remark that the early civilization of the twentieth century ruled Earth when the sun last lay at this particular point in its circumgalactic journey!

We mentioned earlier the so-called Magellanic Clouds. These are other star colonies lying 146,000 light years distant. These systems, of which there are two, are known respectively as the Greater and Lesser Magellanic Clouds. They can be seen only from the Southern Hemisphere, which is rather unfortunate since they are a lovely sight. It is possible to see them even when the moon is full, though of course they can be observed to greater advantage on moonless evenings. The Greater Magellanic Cloud (Nubecula Major) lies in the constellation Dorado, the Lesser Nubecula Minor) in the constellation Tucana. It was thought at one time that these objects represented detached portions of the Milky Way. Now they are known to be separate star systems that are loosely related to our own galaxy.

Having now had a look at the universe in general and our own galaxy in particular, we are in a better position to study in more detail the various bodies that make up the latter.

The major components are obviously the stars, those millions of other suns, some of which adorn our night sky. The briefest examination of the nocturnal scene on a clear, moonless evening should serve to show that stars vary not only in brightness but also in colour. It might be assumed that the brightest are those nearest to us and the faintest are those more remote. This, however, is not so. Much depends on the nature of the star itself. Obviously a very hot and intrinsically bright star, for example, Sirius, which lies at no great stellar distance from the sun, will be a bright object in our night sky. An even hotter and larger one, such as Deneb in the constellation Cygnus, appears almost as bright despite its greater distance from us. On the other hand, the closest star of all, Proxima Centauri (4·2 light years distant), is so faint as to be invisible to the naked eye. The reason for this apparent anomaly should be obvious. Proxima Centauri is a small and relatively cool star. The mighty Deneb, which is 1,600 light years remote, is a flaring celestial beacon whose output of heat and light is 50,000 times as great as that of the sun. We can carry this a stage further by referring to stars that are even mightier than Deneb and are so remote as to be invisible to the unaided eye.

In relation to the planets of the solar system, the sun seems a very large and splendid object, as, of course, it is. Nevertheless, our familiar sun having a diameter close to a million miles is small indeed compared to the incredibly vast Betelgeuse, whose warm, friendly light shines down on us from the magnificent winter constellation of Orion, the Hunter. As these words are being written this lovely star can be seen through the stark silhouette of a leafless tree, its rays flashing above the snow-capped landscape. It is impossible not to marvel at the strange geometry of space that renders a vast sun over 250 times as large as our own no more than a bright point of light. The glow of a minute and lowly firefly would probably appear greater.

Were the great Betelgeuse to occupy the position in space presently held by the sun, none of the innermost planets of the solar system could possibly exist, for the fiery perimeter of this giant red star would reach out to include the orbits of all the planets, even Mars. As is usual in astronomy, this comparison scale works in the other direction, too. Stars are known, the fabulous white dwarfs, whose diameters in many instances are akin to those of mere planets.

In these last paragraphs we have spoken freely of red stars and white stars. This is in accord with those earlier statements in which we said that the colours of the stars vary. A cursory glance at the night sky might give the impression that there is little or no real difference in this respect, that in fact all stars are just yellow points of light. Nothing could be further from the truth.

We can put this to the test quite easily by going out of doors on a clear and preferably moonless evening. Once our eyes have become accustomed to the darkness we can fully appreciate the star-dusted firmament. If it is winter we observe the great constellation of Orion as it strides across the southern sky. There in the top left-hand corner is our old friend Betelgeuse. It is obvious that its light is a gentle orange-red. If now we look towards the bottom right of this constellation we see the beautiful Rigel whose light is unmistakably bluish-white. To the right of Orion, we find the constellation Taurus, the Bull, whose main star, Aldebaran, resembles to some extent the orange-red Betelgeuse. To the lower left of Orion is the constellation Canis Major, the Great Dog, gloriously dominated by what to us is the brightest star in the heavens – Sirius, the legendary Dog Star. The intense whitish light of this star contrasts strongly with that of Betelgeuse and Aldebaran but bears a striking similarity to that of Rigel. Red, orange, bluish-white – are these the only colours to be found? Not at all! By directing our gaze to the high north-east we find the constellation Auriga, the Charioteer, in whose midst shines the lovely yellow Capella.

If the variations of stellar hues are to be seen at their best, however, an astronomical telescope is necessary. By directing the instrument towards certain twin or multiple star systems some truly magnificent sights will be seen. Many of these multiple sun systems display contrasting colours, and seen against the darkness of space they resemble precious stones resting on black velvet. It is possible to mention only a few: the red and green, the cosmic ruby and emerald of the twin suns of Alpha Herculis; the gold and amethyst of 70 Opiuchi; the exquisite blue and green of Alpha Piscium; the striking yellow and blue of Gamma Andromedae; and the magnificent orange, blue, and gold of Zeta Cancri.

That we have chosen to consider the star's colour and true brightness first is no accident, those being the very factors that enable us to sort out the various types of stars and consider the intriguing relationships that they bear to one another. Books on astrophysics and stellar astronomy make considerable use of such terms as *spectral types*, *star classes*, *main sequence*, and so on, and this at first can prove very confusing. Let us see, therefore, if the factors of colour and brightness can be used to make the subject more readily intelligible.

As illustrated in Figure 6b, when a star's absolute (i.e., real) brightness or magnitude is plotted against its colour, some rather interesting and revealing family relationships can at once be established.

The stars having the greatest brightness lie at the top of the graph, and those that are faintest lie at the bottom. Very hot (blue) stars are shown on the left. Stars less hot (white and yellow) are situated in the centre, while the relatively cool ones (orange and red) are on the right.

Now brightness and colour are both related to a star's size. A small star, incapable of a really high degree of heat, is red in colour and quite faint. If the star is somewhat larger, it is hotter and as a consequence appears yellow. If it is much larger, it is extremely hot and bright and burns blue. This, then, provides us with the so-called main sequence of normal, orthodox, 'healthy' stars, the category into which most stars fall. The next two categories contain stars that in many respects must be regarded as abnormal. These stars are giants and super-giants, colossal suns whose sheer dimensions can be described only as awe-inspiring.

If we deal first with the giants, we find ourselves confronted by stars which emit much more light than we would expect on the basis of their respective colours. Since there is a much greater amount of light-emitting surface this is not really too surprising. As a consequence these stars lie well towards the top of the graph, being much higher than their yellow, orange, and red cousins of the main sequence.

What has been said of the giants is true also of the supergiants, but to a much greater degree. Here are stars whose colour range is greater and whose dimensions enter the realm of the fantastic. It is hardly surprising that a 'cool' red star (3,000° C), the diameter of which is 2,000 million miles, should be 10,000 times as bright as the sun. This is a girth more than 2,000 times greater than that of our own luminary. With such a vast extent of light-emitting surface available its brilliance is hardly surprising.

At the foot of our graph is a fourth group or family. This is the group of the renowned white dwarfs, and though their presence may seem anomalous they too can be fitted into the overall scheme. Since they are so small and dim it may seem surprising that they are not red instead of white. The name dwarf is thoroughly merited, for these stars are indeed very small, many of them having diameters similar to that of planets. They represent, in fact, shrunken and dying stars in whose cores the essential nuclear processes have all but ceased. As a consequence there is insufficient energy remaining to withstand the crushing effect of gravity. The density of white dwarfs is almost inconceivably high. It has been estimated that a matchbox filled with the material constituting Sirius's companion white dwarf would weigh many tons. Many physicists now believe that for matter to behave in this way atoms must have been stripped of many of their planetary or orbital electrons. Though small and dying, these stars are far from being cool. In fact they are abnormally hot because their minute dimensions provide an insufficient radiating surface.

From what has been said so far it should not be thought that stars belong to fixed groups from which they can never move. Nothing could be further from the truth. Stars, like human beings, have a definite life cycle. They are born, they mature, they age, and they die in the end. The life cycle of a star is, however, a most complex business, which is even now not completely understood.

If we return to our graph and consider the particular case of just one star, our own sun, some understanding of what is

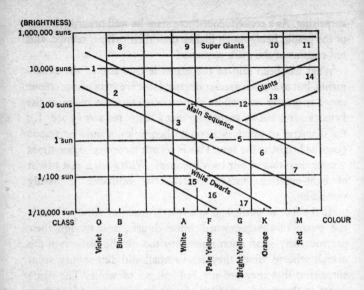

Key:

	Diameter (Millions of miles)	Surface Temp. (°C)	Class	
1. Violet	8	35,000	O	
2. Blue	10	25,000	B	
3. White	1¾	11,000	A	(Main
4. Pale yellow	1½	7,000	F	Sequence)
5. Bright yellow	1	6,000	G	
6. Orange	¾	5,000	K	
7. Red	½	3,000	M	
8. Blue	30	25,000	B	(Super-
9. White	70	10,000	A	Giants)
10. Orange	1,000	3,500	K	
11. Red	2,000	3,000	M	
12. Yellow	7	5,000	G	
13. Orange	17	4,000	K	(Giants)
14. Red	60	3,000	M	
15. White	Of the order	Temperatures		
16. White	of 5,000 miles	Uncertain		(White
17. White				Dwarfs)

Figure 6b

involved may be gained. If we look at the graph in Figure 6c, a reproduction of the graph in Figure 6b, we can see the life path of the sun represented by a dotted line. Very long ago – it may have been as far back as 5,000 million years – the sun condensed out of a gaseous nebula into a fairly insignificant red star. (See

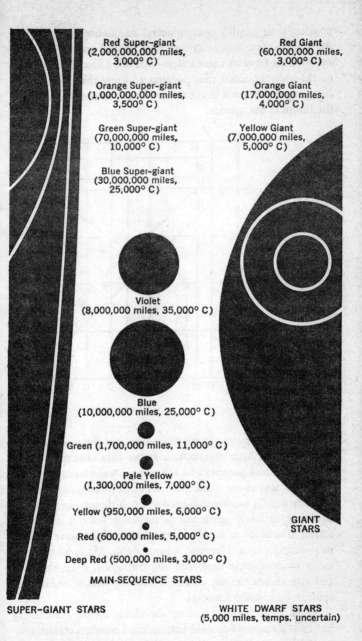

Chart illustrating sizes of the stars, from *How We Will Reach the Stars* by J. W. Macvey, (Collier–Macmillan, 1970).

position 1 on graph.) Quite quickly, astronomically speaking, the great nuclear fires at its core were 'stoked up' and the sun assumed the form we know today – that of a Class G yellow star on the main sequence. (See position 2.) From this point on we have a special interest in the matter for the future of our sun is the future of our species.

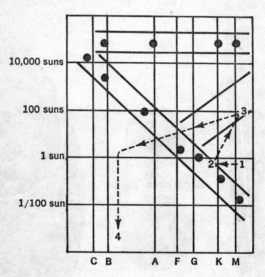

Figure 6c

Our sun is expected to remain unchanged for another 5,000 million years. After this certain rather dire changes become imminent. From position 2 on the graph it will move to position 3, the domain of the giant stars. In other words, our sun will swell up until it has become a red giant. Although the period of this very turbulent phase will be relatively brief, it will prove utterly disastrous to all the inner planets of the solar system, including our own. Mercury, Venus, Earth, Mars, all will certainly be completely destroyed and consumed. Even the aloof and magnificent Jupiter, with its splendid system of moons, will come in for a severe roasting!

Throughout recorded history the theme of the world's end has been a source of morbid fascination. Countless cranks, only to be frustrated, have predicted the exact moment of cosmic doom. A legion of science-fiction writers have described in lurid, horrifying detail the coming of this awful, final day. Their fertile pens have shown us Earth being destroyed by comets, shattered by wandering asteroids, and frozen into extinction by the sun's death. On more than one occasion we have, in imagination, been witnesses of the last dawn, spectators of a sky already filled with the strange onrushing planet that within hours will fuse with Earth in a cataclysmic and fiery embrace. It is only natural for us to wonder at times just how and when the life of our small world will end. Here, then, is the answer: it will quite literally be consumed by fire but fortunately at a point so remote in time that it need not concern us. Life may indeed have disappeared from Earth by then or have undergone so great a measure of evolutionary change as to render it totally unrecognizable. But if life, a life like ours, does remain, it seems fairly safe to say that it will not see the fantastic roseate sunrise of the last day – the day on which the great red outer shell of the sun sweeps inexorably towards our doomed planet. Soaring temperatures, melting mountains, and boiling oceans will long since have rendered Earth as sterile as the day before the first primeval speck of life stirred on the sea bed!

Following the state of stardom of red giants, our now unfamiliar sun will pursue the arrowed course plotted on the graph. From a red giant it will shrink first to an orange giant and then to a much smaller yellow one. Finally it will reach the point of senility and the status of a small but hot white dwarf.

Although we obviously cannot go deeply into the astrophysics of these peculiar changes, we cannot in fairness pass on without a modicum of explanation. We will try, therefore, to outline very briefly the probable reason for these changes.

In the beginning we must visualize a cloud of dust and gas containing centres of high density around which it begins to contract. If there is just a single such centre only a single star will result. If, however, there are two or more centres, then a binary or multiple system can be expected. Contraction on the part of this stellar embryo or protostar results in atomic collisions within the compressed matter and thus to the release of energy. In time this heat becomes so intense that the familiar hydrogen-to-helium fusion process begins to occur. Given adequate time this energy proves sufficient to oppose and ultimately to overcome the contraction brought about by gravity. At this point our star has become stable and reached a mature state. It could in fact be said that it has come of age.

If the star is very large and massive the gravitational effects will be extremely powerful and quick. Consequently, the amount of initial heat generated will be extremely high and will lead to the emission of vast quantities of fusion energy. This is an essential requirement, since balance must be effected. Conversely, a small star will be required to produce much less fusion energy.

Our star is now in the region of main sequence, as illustrated in the diagram in Figure 6a. The actual point it occupies is really dictated by its mass, which is responsible for a particular value of gravity/fusion balance. Thus we find that the so-called main sequence consists of a complete range of star types. If we start with the large, intensely hot and fast-burning blue stars we find that the sequence goes on to embrace the yellow, much less massive, cooler, and slower-burning stars and then the small, red, slow-burning 'cool' variety.

A star spends the greater portion of its existence in the main sequence, although the actual period does depend on its nature. Thus if it is a massive, hot, blue star the time it will spend in the main sequence is unlikely to exceed a few hundred thousand years. A medium-sized yellow star like the sun can be expected to remain there for a few thousand million years, whereas a small, cool, red star will exist in the main sequence for several hundred thousand million years.

Just precisely what does take place at the close of a star's main-sequence life is not as yet clearly understood, even though astrophysicists now believe that once approximately 10 per cent of a star's hydrogen has been converted, an accumulation of helium ash begins to form at its core. Around this core the fusion process continues unabated, and as the amount of ash increases the additional weight causes the core to contract. Atoms are thereby compressed causing electrons to be squeezed from their orbits. As before, this leads to a massive release of gravitational energy in the form of heat. This heat, pouring copiously from the core, greatly accelerates the pace of the fusion reactions already in progress around it. This in turn causes a tremendous outpouring of energy from the star whose outer regions consequently become violently disturbed. Not only does the brightness of the star increase, but it also begins to expand to a fantastic extent. Since its outer shells are thus further removed from the rampaging nuclear 'fires' deep within, their temperature begins to fall, and so the star assumes the typical red or orange bloated appearance of a member of the giant or supergiant series. Its actual position in either of these categories depends very largely on its original positions of main sequence. A star many times more massive than the sun will become a red super-giant like Antares or Betelgeuse, whereas a star more akin to the sun can be expected to attain the status of a much less swollen red or orange giant.

As might be expected, the accumulation of helium ash piles up more quickly now that the star is virtually running wild. This further increases the gravitational effects and, in turn, the rate of fusion. Here we have a vicious circle in the fullest sense, for the overall result of all this is that helium ash accumulates at an even higher rate. Our star is now, in fact, out of control!

When a star has reached this condition it is obvious that further changes of a radical kind must be anticipated. Therefore, it will probably come as no great surprise to learn that our red giant does an about-turn and eventually begins to burn blue once more.

Where a yellow solar-type star is concerned, this stage is thought to proceed along the following lines. After some 40 per cent of the hydrogen has been used up, the core of helium ash contracts under gravity to such an extent that temperatures in excess of 100 million degrees are produced. Under these conditions our hitherto inert and long-suffering helium ash decides that at last it has had enough and that the time is ripe for it to take a more active part in the proceedings. This it does by emulating hydrogen, its predecessor, and changing into something else, in this instance the elements carbon, oxygen, and neon. This in turn results in gamma radiation, which is able to convert the inert gas of the inner core to a very active variety of gas.

At this point the star has reached a most unstable state in which the core may well explode. This convulsion, though incapable of shattering the star completely, does nevertheless achieve some most remarkable results. The vast concentric shells of matter absorb not only the shock of the detonation but also the heat associated with it. Once the dangerously 'unhealthy' core of the star has been eliminated it might seem that the star should return after a period of convalescence to a state of complete stellar health. This is precisely what does *not* happen, although the mechanism of what does actually take place is far from clear. A host of new nuclear reactions is apparently initiated in the star, which may now be regarded as a vast series of concentric shells. In each of these shells a nuclear process is under way, which provides a supply of ash for the process going on in the one immediately within.

Our star is fast heading now towards the state of white dwarfdom. In fact, it is dying. Soon, on the stellar scale, the remainder of the consumable nuclear fuel will be expended. Once this supply is gone, the star must contract under gravity to the point where any further contraction is effectively resisted by atomic nuclei. The star will then have become a genuine white dwarf. No further disruptive or cataclysmic spasms will cloud its twilight aeons. Over millions of years it will continue to cool until the time must come when it is as inert and cold as the terrible vacuum of space surrounding it. So passes the glory of a star!

On the whole, stars are fairly stable bodies within the limits already specified. Although their swelling to gianthood must inevitably prove catastrophic to any planets orbiting nearby, this phase is nevertheless one that follows fairly gradually. There are, however, exceptions to most rules, and this is certainly applicable in the case of stars since there are those that from time to time behave in a most unorthodox and indeed disastrous manner.

Giants and super-giants, although admittedly abnormal in the pace at which they use up their substance, are nevertheless doing this in an essentially steady way. The same can hardly be said for those flaring celestial beacons known to astronomers as *novae*. The term novae really means 'new stars', but this is a distinct misnomer and constitutes a relic from a much earlier era in astronomy. Occasionally a star seems to appear at a point in the heavens where apparently none existed before. Sometimes these occurrences are marked by tremendous brilliance. Indeed, on occasion it is as if a new Sirius had suddenly appeared in the sky. What of course has happened is that a hitherto distant and relatively insignificant star, either at the threshold of naked-eye visibility or beyond, has suffered a unique and cataclysmic convulsion, making it much more brilliant. Before the invention of the telescope and the realization of some stellar processes, it was quite natural to attribute these occurrences to the birth of a new star. Unfortunately the records do not show just how this near-miracle was supposed to have taken place. We may smile at the seeming naïveté of the ancient sky-watchers but we should certainly not be scornful. In their way these people did much for astronomy, and after all it is not impossible that astrophysicists a few centuries to come may regard some of our theories with equal amusement.

There are several different types of novae. Some are merely novae-like and flare up at irregular intervals but remain as 'going concerns' after the outburst. For long periods they are dormant and apparently normal. In some cases they emit great opalescent shells of gas that, blown outwards by the violence of the reaction, remain illuminated by the light of the parent stars. The famous and extremely beautiful Dumbbell Nebula in the constellation Vulpecula is a typical example. The parent body in this case is a small, intensely hot, blue star whose ultraviolet rays cause the gas of the nebula to fluoresce in the most colourful manner. The impression is of a great soap bubble floating serenely in the depths of space. The colours range from blue at the centre to a delicate shade of red at the edges. Such glowing spheres of fluorescent gas can have diameters of several billions of miles, and though they may seem immobile these spheres are probably being propelled outwards at speeds of anything up to 100,000 miles per hour.

Novae and supernovae are cataclysmic stellar convulsions that are virtually impossible to describe. These are stars that have quite simply exploded, leaving only shattered debris behind them. The 'wreckage' of several such cosmic detonations may be found at a number of points in our night sky. One of the best examples is the Crab Nebula in the constellation of Taurus, the Bull. This nebula represents the aftermath of a supernova that shone in our skies in A.D. 1054. According to old records this stupendous stellar blast was of such brilliance that for two years it shone *by day* as well as by night! Here is the story of disaster

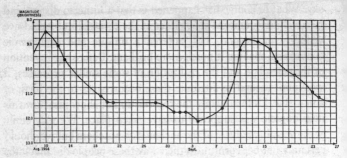

Figure 6d. Light curve of SS Cygni (August 1964–November 1964).

written quite literally in fire on the scroll of the heavens. This was a star, a sun, in many respects not wholly unlike our own, that was suddenly rent asunder. Though it lay 3,300 light years from Earth neither this great distance nor the brilliance of our sun could dim the great convulsive fires of its death agonies.

The most common of the unstable stars are those generally grouped under the collective title of *variable stars*. This is a category that includes many different types of star, but all have one feature in common, and that is their varying heat and light output. These variations in no small number of cases are extremely regular and may even be timed to the minute. Others again are much more irregular. This conveniently provides us with two fundamental subdivisions – regular and irregular variables. A further element of subdivision is still called for, however. Regular variables may have long or short cycles. In contrast, irregular variables are likely to vary in a number of ways. The latter category is especially fascinating and constitutes an excellent subject for the serious amateur observer. The main element of fascination lies in the fact that from night to night some significant, even spectacular, change may take place in the brightness of the star being watched. Some increases in brilliance come along much more rapidly than others with even the extent of the rise differing.

One of the most interesting of these irregular variables lies in the lovely northern constellation of Cygnus, the Swan. This is a large and rather striking constellation that is with us for most of the year in various positions in the sky. Around the end of May it can be seen rising in the north-east towards midnight. During

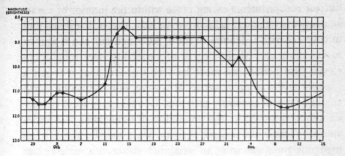

late summer and autumn it dominates the sky, and in the early evening at Christmas it is almost directly overhead. Thereafter it gradually sinks into the haze and murk of the low north-west where it remains until April. This constellation contains the great, hot blue star Deneb as well as a number of well-known and important variable stars. Among these variables is one known to astronomers as SS Cygni. This star might be described as one that is fairly regular in its cycle of irregularities! Its periods of brilliance generally follow one another at an interval of about fifty days. At times it rises to a brilliance of 8·1, while at other times its peak brilliance may only amount to around 9·0. Perhaps for the benefit of readers unfamiliar with the terminology we should explain that the extent of a star's apparent brightness is referred to as its magnitude. If a star is of magnitude 1 or less, it is a very bright one; if the magnitude lies between 1 and 2 it is somewhat less bright and so on. Beyond a magnitude of 6, a star is invisible to the unaided eye. SS Cygni can have a range between a maximum of 8·1 (peak brilliance) and a minimum of 12·5. It is, therefore, a 'telescopic' variable since it never comes within a range visible to the naked eye. The graph shown in Figure 6d (taken from the writer's own observatory records) shows how this star varied in brightness between August 1964 and November 1964.

Briefly, then, variable stars are stars whose output of energy is not constant. The cycle of variation can be long, short, or irregular, and the extent of the variations may also be different. The reason for this peculiar and erratic behaviour is even now not fully understood. It may well be due to drastic compensating nuclear readjustments taking place within the interior of a star. In many cases the star is known to pulsate, that is, to expand and to contract with a definite rhythm. Although our sun is known to exhibit a very slight degree of variability over roughly an eleven-year period, we must be very thankful that it is not a variable in the accepted sense. If it were, our climate on Earth would be one of appalling extremes. Coupled with these climactic extremes would be the possibility of devastating and widespread volcanic and seismic activity. Days may lie ahead when incoming messages from an inhabited planet of a variable star will tell us vividly of life under the terrible rays of an erratic sun!

Before looking at star clusters and gaseous nebulae, there is one other type of star that we should consider. This is the binary or double star system, or, if there are more than two components, the multiple star system. We know that our own sun is merely another star, and familiarity with this idea might easily lead us to believe that all other stars are akin to our sun apart from essential differences in temperatures, density, size, and colour. Many of these other suns are, however, systems that have two or more components and as such are very different indeed from our own life-giving luminary. These systems are very common, and a small astronomical telescope will reveal literally scores of these lovely stellar groups.

Earlier in this chapter we mentioned the sheer beauty of stars showing contrasting colours. Although it is unfortunately quite impossible to obtain a close view of any of these systems, we can perhaps still visualize something of this sheer spectacle were we to be suspended in space only a few million miles from one of them. Let us for a few moments imagine that a miraculous starship of the distant future has carried us close to the amazing twin-sun system of U Cephei. Beyond in the limitless immensity we see the usual panoply of star-powdered splendour. Directly ahead lies a sight that would impress forever even the least impressionable. Two immense suns, one golden yellow and the other pale blue, sweep through the heavens together. Because of the violent gravitational pull exerted on one another by the two stars, each has become egg-shaped. Streams of incandescent orange-red gas flow between and around the two in a way that is both terrifying and awesomely beautiful. It is the sort of sight that no mortal artist could ever hope to portray, one that no brush could ever capture. Here is an epic example of the immense and almost limitless forces at work in the physical universe.

Just why should binary and multiple star systems exist? Like many of the questions in astronomy this is not an easy one to answer. One theory, which today finds a fair measure of acceptance, envisages gaseous nebulae condensing around more than one point. If there are two such points a binary system results, if more than two a multiple system evolves.

If intelligible signals ever reach us from inhabited worlds of multiple star systems they can hardly fail to be of unique interest. At present we can only imagine the type of spectacle likely to be afforded by a sky containing more than one sun. How intensely fascinating it would be to learn of the double-shadow effect brought about by twin suns or the fantastic kaleidoscopes caused by the rising or the setting of triple suns of different colours.

Finally in this brief general summary of our universe we come to gaseous nebulae and star clusters. Gaseous nebulae have been described in a variety of colourful and picturesque ways. The famous M42, or Great Nebula in Orion, has been called 'a crucible of stars', and the truly exquisite Bridal Veil Nebula in Cygnus has been referred to as 'the stuff whereof the stars are made'. Both of these romanticized descriptions reveal a single basic truth: that it is from such vast elemental clouds of glowing hydrogen that stars are created. When, therefore, astronomers focus their great telescopes on these lovely and often ethereal-looking objects they are in effect seeing the raw material of future stars and even in some cases the actual birth of a star itself. Like all things astronomical, of course, the birth process is of fantastic length by mortal standards. To great wheeling galaxies of stars and glowing clouds of primeval gas the human lifetime is but the minutest fraction of a second.

We will now concern ourselves with star clusters, which, as their name implies, are concentrated groupings of stars that are classified as either 'open' or 'globular'. It is believed that both types owe their existence to the same process that may have led to the creation of multiple star systems, that is, condensation from hydrogen clouds. In this instance, of course, the clouds were extensively greater. Even though the stars in a cluster may apparently lie very close to one another this is merely another of the visual effects of immense distance. The sun and most of the stars we see on a clear moonless night belong to our own local cluster, yet the nearest star to the sun is Proxima Centauri at a distance of 4·2 light years. Thus if all stars in such a cluster originated within a single hydrogen cloud, then that cloud must have been of vast dimensions.

Open clusters are the more common of the two types and relatively speaking are also much nearer to us. Several are indeed very well-known and are plainly visible to the naked eye. Probably the most familiar is that compact little group known as the Pleiades, or Seven Sisters, in the constellation of Taurus, the Bull. The rise of the Pleiades in the east during the late evening is a sign of approaching winter, and, to all who are watchers of the sky, the coming of this little group is like the return of an old friend. The name Seven Sisters is today something of a misnomer. Generally, even people with the keenest eyesight see only six stars. There are reports of people seeing up to a dozen stars with the unaided eye, but this seems rather improbable. The apparent compactness of the group should not be misinterpreted. The group is about 300 light years remote, and its real diameter is in the region of 15 light years. Thus, even though they may appear close together, the individual members of this cluster are in fact widely separated. Nevertheless there exists a distinct physical relationship between them, for they are known to be travelling together through space like a family of wild geese flying through our earthly skies.

It has been estimated that if our sun were to lie in the middle of this lovely group our night sky would contain many stars shining with an even greater lustre than the familiar, flashing Sirius. And so if ever a signal should originate from somewhere within the confines of the particular 'corner' of space occupied by the Pleiades, it might tell us of the wonders and beauty of this alien yet fascinating nocturnal scene.

Globular clusters, apart from being much more remote, contain many more stars than their counterparts, the open clusters. They lie thousands of light years distant and constitute what we might loosely describe as the outer, spherical 'shell' of the Milky Way galaxy. They represent, in fact, a sort of galactic last frontier, for beyond these clusters lie only the incalculable, reason-destroying deeps that separate universe from universe. If beings from another galaxy were ever to succeed in crossing this most terrifying abyss, it is these globular star clusters that they would first encounter as they approached the Milky Way. Creatures on worlds within these peculiar stellar colonies could, in their electronic narratives, tell us of life on the fringes of our universe, of existence on the very shores of the vast intergalactic ocean!

Few globular clusters are visible to the unaided eye. One of the most notable exceptions is the well-known M13 in the constellation Hercules, which appears as a faint, hazy patch of light. Through a powerful telescope this is a most beautiful sight.

So much, then, for the Milky Way, that great galaxy of which our solar system forms such an infinitesimal part. We have touched on only its main features and these quite briefly. Nevertheless this may appear to be a considerable digression from our main theme. If, however, we are to study the possibilities and the enthralling potential of interstellar communication it is essential that we have some idea of the great stage upon which this exciting and fascinating drama will be played. We must know something about the sources of the radiation that may whisper one day in our terrestrial antennae and of the desolate, yawning emptiness of the medium through which this will 'crawl' with the speed of light. If we learn of a civilization whose message reaches us after a thousand-year journey, it is desirable that we possess some awareness of the particular cosmic environment enjoyed (or once enjoyed) by that civilization. We should never be unmindful that messages that come from the stellar depths must also be messages out of the past. When we survey the stars we see them as they once were; when we listen to the signals of their civilizations we hear those as they once were. We see the far star mists as they were perhaps 100,000 years ago – the nearest stars, as many were a few years ago. So, too, must it be with radio signals. Some may have originated in the incredibly remote past, others in a past considerably less remote, and some from a past that is very recent. It is never possible for us to view a star as it is at this moment.

We can never learn of an alien civilization as it is now. This is one of the strange truths of interstellar communication, one of its most fundamental ironies – and there is no way round it! Amid these great cosmic deeps, light is not the instantaneous light of Earth, and time is not the dimensionless entity of this little planet.

4 The Star Worlds

IT IS obvious that we can communicate with other galactic races only if such races exist. If these races do exist it is equally plain that there must be planets for them to exist upon. It is desirable, then, to review briefly the question of extrasolar planetary formation. Is there a real and valid basis for the belief that the sun is not unique by virtue of its planetary retinue?

From a fairly early stage in modern astronomy, attempts were made to formulate a theory that might satisfactorily explain the origin of the solar system. Ideas on the subject changed considerably over the years, and it is perhaps ironic to record that the theory presently finding acceptance is in many respects akin to one first suggested over two hundred years ago and subsequently discarded.

It must be obvious to any rationally minded person that the solar system is not merely a chance collection of bodies. If we include the very large and continually growing number of asteroids, or minor planets, we find that there are approximately 1,500 planets in our solar system. All these planets revolve around the sun in a uniform direction, their orbits lying in virtually the same plane. If we ignore Pluto and a few asteroids that cross the orbits of other planets, it is a fact that all the orbits are very nearly circular. Moreover, the sun is known to rotate in the same direction as that in which the planets revolve, the orbits of the latter lying in virtually the plane of the sun's equator. The orbits of the planetary satellites are, by and large, circular and lie nearly in the planes of their respective planets' equators.

Had the solar system been a chance assemblage it is most improbable that all this would have been so. The odds against such a high degree of coincidence are immense, and it is much more rational to attribute these features to a common cause that nevertheless must have been something quite grand on the cosmic scale.

The first real hypothesis relating to the formation of the solar system must be attributed to the German philosopher Immanuel Kant, who, in 1755, postulated that the material represented by the sun, planets, satellites, asteroids, and so on had once constituted part of a great diffuse cloud or nebula in which the dominant tendency was for the heavier elements to fall towards the centre. This process was, however, opposed by the expansion of the gas within the nebula. As a consequence, lateral motions were initiated that imparted rotation to the entire nebula.

It was later shown that rotation of the nebula could not have been brought about in this way because such a process would have been contrary to an important physical principle known as the conservation of angular momentum. The angular momentum of a body may be defined as a measure of the total rotational motion that body happens to possess. Kant believed that from a collection of localized rotations there must develop a general rotation. Unfortunately Kant's logic was faulty, for the resultant general rotation of the nebula, which he postulated, would require to have been produced by local rotations working *in different directions*. Initially, angular momentum was zero, and the general cancellation effect of these local rotations upon each other could only have been to keep this at zero.

Towards the close of the eighteenth century the celebrated French mathematician Laplace suggested that the original nebula had in fact already possessed a slight amount of rotation. As it cooled it contracted, and so its density increased. In order to conserve angular momentum the rate of rotation also increased. Thus a process was initiated that led eventually to equilibrium between gravitational pull and centrifugal force. As a result, a circular belt of gaseous matter was flung outwards from the equatorial regions of the nebula. The nebula continued to contract, and in due course the process was repeated. This took place several times until eventually the nebula was surrounded by several gaseous belts. The latter condensed accordingly to form the future planets of the solar system, which, still in a gaseous state, went through a similar process on a reduced scale to produce the various satellites.

For a time this theory was able to account quite satisfactorily for the existence and position of the various major components of the solar system. Later, however, difficulties began to arise. Laplace had unfortunately been rather arbitrary in his assumption that each individual ring of matter thrown out by the nebula must condense into a planet, and he had made no real effort to prove his dictum in mathematical terms. British physicist and mathematician James Clerk-Maxwell was able to show that a single body could not possibly coalesce from such an ejected belt of hot gaseous matter. It was, he argued, considerably more probable that a ring of this nature would continue to exist in the form of a stable system of very small bodies similar in many respects to the famous rings of Saturn. It could also be shown that of the total angular momentum contained by the solar system, including the sun, the planet Jupiter was responsible for over 50 per cent. Indeed the four major planets, Jupiter, Saturn, Uranus, and Neptune, account for no less than 98 per cent of the total, leaving only 2 per cent to be contributed by the sun and the lesser planets. This was indeed a peculiar state of affairs, for the four giant planets accounted at the same time for only 1/700 of the solar system's total mass. Since the angular momentum of the system must be equal to that of the original nebula out of which allegedly it was created, some reason must be found for such a high proportion of that momentum having been bestowed on a relatively small portion of the system. Laplace envisaged a contraction process whereby angular momentum was neither gained nor lost, and so the highly anomalous distribution within the solar system could hardly have been brought about by the sequence of events he had suggested.

It is possible to make a reasonably satisfactory case by assuming that the major planets obtained their excessive degree of angular momentum by virtue of an external source totally unassociated either with the sun or original nebula. Such an assumption leads directly to the famous hypothesis of the passing star. Briefly, this hypothesis suggests that at some remote epoch in the sun's past another star made a very close approach to it. In the process a gigantic protuberance was raised on the surface of the sun (and presumably on that of the other star as well). As the distance between the two stars steadily shrank, the greater these vast tidal phenomena grew. As the star passed close to the sun a point was ultimately reached when the former's gravitational pull finally overcame that of the sun. At this stage a long cigar-shaped filament of matter broke away from the crest of the great tidal mass raised on the sun. Of this matter some was captured and retained by the passing star and some was lost in space, while the rest remained under the influence of the sun to condense eventually into the planets of the solar system.

A theory of this nature certainly accounts quite satisfactorily for the unequal distribution of angular momentum within the solar system. The long filament of matter pointed towards the star could be expected to receive some of the latter's angular momentum. Such a concept also serves to explain why planets and their satellites orbit in virtually a common plane and why, too, they display a uniform direction of revolution.

The theory of the passing star became in time the fountain from which several variations were to spring. Towards the close of the nineteenth century, T. Moulton and Thomas C. Chamberlin used the original postulate as a basis for their celebrated planetesimal hypothesis. The sun, the surface of which is in a state of continual turbulence, was again regarded as having been closely approached by another star. The masses of incandescent matter that the sun is always hurling outwards were acted upon by the passing star so that instead of falling back into the sun they were permanently parted from it. From these masses, according to the hypothesis, the four major planets were born. Similar ejections from the opposite side of the sun, where the gravitational pull of the other star was considerably less, produced the smaller planets, one of which eventually became Earth. The ejected material was regarded as having cooled rapidly, forming many small liquid bodies each moving round the sun in an independent orbit. These bodies were termed planetesimals, and their distribution was by no means uniform, being more concentrated in certain regions than in others. Where these concentrations were most dense they tended to coalesce into large single solid bodies that began to act as a sort of focal point for others. Thus the planets were born. The asteroids, according to this theory, were seen as planetesimals that in the absence of a large aggregation were allowed to retain their individual existence. Satellites, on the other hand, were regarded as planetesimals possessing sufficient velocity to prevent their being swept up by the larger units but insufficient velocity to escape entirely.

Not all of the material ejected by the sun came completely under the influence of the passing star or remained in space to form planets. Some, it was supposed, fell back into the sun, and since this momentum in turn was imparted to the sun, the planets and the sun came to possess the same direction of rotation.

The fundamental flaw inherent in this ingenious hypothesis lay in the fact that the sun is unable to eject material of a mass sufficient to form bodies like planets. The force largely responsible for this inability is the pressure of the sun's radiation, which is sufficient only to impart high velocities to particles having molecular dimensions.

At this point in the proceedings, a British cosmologist, the late Sir James Jeans, appeared on the scene with his renowned theory of tidal ejection. This was merely a new variant of the passing star theme, or more correctly it was the Moulton–Chamberlin modification now in turn modified.

Jeans showed that the sun itself must strongly influence the effects brought about by the close passage of another star. Were the sun of uniform density and incompressible it would act as a solid body, in which case the close approach of another star could result only in the former sun's total disruption into very large fragments. If on the other hand the density of the sun showed gradations that increased towards the centre, such fragmentary products would be small and the sun's mass therefore only barely affected. In such circumstances matter pulled out by the gravitational attraction of the other star would have originated in the outer layers of low density. Since we are aware that the density of the sun increases in the direction of its core, it is this second case that is applicable.

Therefore, a tremendous bulge would first be raised upon the surface of the sun that would very quickly grow into an odd-shaped protuberance. From the outer tip of this protuberance a long cigar-shaped jet of incandescent gas would stream off into space. It is unlikely that this jet would be of uniform density, and there would consequently exist a distinct tendency for the material to condense around centres of high density. Thus the vast jet of fiery matter many millions of miles in length would break up to form a number of separate bodies. Once the other star had proceeded on its way, these planets would take up their independent orbits around the sun.

It is reasonable to assume that the greatest outrush of matter would occur when the passing star was at its nearest to the sun and the least during the periods of its approach and departure. Thus we could expect the stream of matter to be thickest in the centre and narrowest at its extremities. The larger planets would therefore coalesce in the centre and the smaller towards its tapering ends. This, of course, is precisely what we find; at the centre of the sun's planetary family lie the largest planets, Jupiter and Saturn, and at either end lie Mercury and Pluto, which significantly are two of the smallest planets. Intermediate positions are also occupied by planets that have appropriate sizes. This does not mean that there are no anomalies. Mars is very much on the small side but the gap between it and Jupiter is too extensive. This, however, is the region of the minor planets, the so-called asteroid belt. There are those who believe that this may originally have been occupied by a large single planet that for some obscure reason was disrupted into innumerable small fragments. Another possibility is that lower density in this region produced only small centres of condensation.

The planetesimal hypothesis suggested that matter not swept up by the embryo planets remained instead as pockets of gaseous material and small solid particles, a belief that also finds expression in the Jeans philosophy. The presence of satellites is attributed to a mechanism not unlike the one that created the planets, but it is carried out on a much smaller scale. Sun and star created tides on the primordial planets, and by an analogous process these condensed to form satellites.

At first the Jeans theory looks quite palatable, but a more analytical examination soon reveals serious shortcomings.

Originally it was supposed that solar rotation was the result of ejected matter falling back into the sun and that planetary rotation was derived in an analogous manner. Let us, however, delve a little more deeply into the implications. Jupiter is known to possess a very rapid rate of rotation, and it has been estimated that to produce this rate about $1/15$ of its mass would have to fall back. This represents an inordinately high proportion, about equal in mass to 400 times the aggregate mass of the planet's twelve satellites. This seems a hardly credible state of affairs. The same can be said in reference to the three other large planets, Saturn, Uranus, and Neptune. Mars possesses two extremely minute satellites while Venus and Mercury have no satellites at all. In these instances are we to assume that ejected matter fell back almost in its entirety?

In time it became clear that the validity of the entire concept of the passing star was doubtful. Nevertheless this did not deter its protagonists from postulating further extensions of the theme in an attempt to retrieve the situation. These only served to render the theory more untenable still. Some sought to suggest that instead of merely making a close approach to the sun, the errant star actually collided with it or at least struck it a glancing blow.

Let us examine this idea for a moment or two. As sun and star drew steadily closer their velocities increased. At the moment of impact these velocities had become very considerable indeed. As before, vast tongues of incandescent matter sprang from the surface of each body and began to break free just prior to collision. The actual impact resulted in a certain intimate mixing between the outermost shells of the sun and the star. The dense central mass of the star then swung about the sun before continuing on into space. This mutual layer was not only extremely hot but it also possessed a very rapid rate of rotation due to the shearing motion of the two stars. In this way a rapidly rotating filament of material, roughly analogous to that postulated in the Jeans concept, was produced between the sun and the intruder.

Although this may answer, or at least appear to answer, one question, it also poses several others, mostly of an awkward nature. One of these relates to the temperature of the gaseous stream so produced, which it is estimated would have been as high as 10 million degrees. At such an extremely high temperature dissipation of the material into space seems a more reasonable prospect than its condensation into planets.

Serious discrepancies also begin to arise when we look again at the question of angular momentum. A mere close approach by the intruder star would have been responsible for imparting less angular momentum than is contained by the entire solar system at the present time. The position is only worsened by postulating an actual collision, for this would have imparted even less.

Yet another variant of the theme was introduced by the suggestion that when the intruder came along our sun was a member of a binary star system. Such systems are of course not uncommon. The sun's erstwhile companion, we are told, was slightly smaller than the sun from which it was separated by a distance roughly equal to that between the sun and Saturn today. The intruder, sweeping in from interstellar space, collided catastrophically with our sun's alleged companion. Most of the resulting mass of fragmentary material was captured and borne away by the stellar juggernaut leaving behind only debris that in time became the planets.

The most glaring weakness in this theme is that such a cataclysmic event would probably have been unable to produce a system of planets all orbiting the sun in more or less the same plane. Nevertheless it is a modification that does admittedly tend to eliminate the earlier obstacles of rapid planetary rotation and high angular momentum. The rotation in such circumstances could be attributed directly to the impact, and the angular momentum could be said to have been possessed initially by the sun's unfortunate consort. In nearly all other respects the result of this modification was merely to heap difficulty upon difficulty for it raised a host of new and irreconcilable facts.

In the end the passing-star concept faded into oblivion. It had been a sincere, ingenious, and colourful attempt to explain the origin of the solar system, but from its inception it suffered from what, in retrospect, is an appalling artificiality. Nevertheless over many years it enjoyed considerable, indeed almost fanatical support. Just why the era of anthropocentrism should have endured so long is still difficult to explain.

Had this concept been finally endorsed, the implications so far as stellar worlds are concerned would have been decisive. In the circumstances other solar systems *might* have come into existence, but like our own they would have ranked as galactic freaks because they would have been fantastically rare.

The fundamental reason for this state of affairs lies in the vast and virtually incomprehensible distances that separate star from star. The possibilities of a stellar collision or even of a near approach between two stars are therefore very slim indeed. This single incontrovertible fact probably strikes more deeply at the passing-star theory than any other fact.

Ironically, Jeans himself worked out the chances of stellar collision, and in so doing he may have unwittingly helped to undermine his own theory. He believed that the likelihood of one star ever being in collision with another was of the order of *one* in 600,000 million, million years. Later estimates have been less conservative, putting the chances of collision or near approach somewhat lower. A recent appraisal of the position suggests that only one star in 50 million is likely in its lifetime to approach or collide with another.

With the demise of the various theories of the passing star it became feasible once again to consider seriously the existence of worlds lying deep in space among the stars. There still of course remained one essential condition to be met: a theory that adequately explained the origin of the solar system must be fully applicable to other star systems as well.

The theory that ultimately found acceptance is essentially a development of the earlier propositions of Kant and Laplace. Whereas the passing-star concept contained certain merit, it seems eminently more reasonable to look for a theory based on the cumulative action of natural events rather than on chance, cataclysmic catastrophe.

In this respect the present theory developed by C. von Weizsäcker is fully acceptable and shows promise in a number of ways. It envisages an early sun surrounded entirely by a shell or envelope of gas and dust with a mass equal to about one-tenth that of the sun itself. Dimensionally this is regarded as being comparable to the present-day solar system with a constitution akin to that of the sun, namely, hydrogen and helium. This envelope, it is thought, gradually assumed the form of a disc girdling the sun. There then followed a slow transfer of angular momentum from the inner to the outer regions, from which it was lost as gas escaped into space. The succeeding stage is seen as one in which the material of the disc formed localized vortices of varying size. These, however, were in an almost continual state of flux, forming, breaking up, then re-forming. Eventually this great disc of semi-independent vortices broke up permanently into individual clouds or protoplanets, the masses of which were dependent to a large extent on local condensations of gaseous matter. Further condensations in the central regions of the protoplanets led subsequently to the formation of the planets proper and their satellites.

Probable Sequence of Events leading to Solar System

PRIMORDIAL NEBULA
↓
SUN SURROUNDED BY CLOUD

(This cloud may have been uniform *or* non-uniform in constitution.)

UNIFORM CLOUD   NON-UNIFORM CLOUD

Distribution of elements in inner and outer planets indicates that fractionation has taken place (due probably to sun's radiation). Thus more volatile elements form *outer* portion of cloud.

Eventual condensation of bodies from sun-cloud system. This may have led directly to present planets or to transitional series of protoplanets.

PRESENT PLANETS   PROTOPLANETS

After due process of cooling

Perhaps three in number, which eventually broke up to form present planets. If these possessed a layered structure, composition of resultant planets would be influenced. This is a likely premise in view of the fact that the Earth, moon, and Mars (thought to have originated in the same protoplanet) have differing densities. Meteorites (originating in one of the other protoplanets) also vary in respect of density and composition.

This is a greatly simplified version of what in fact must have been an exceedingly complex process. Already several modifications have been proposed but these must rank more as refinements than attempts to remove anomalies. On the whole this theory represents a practical, sensible, and essentially plausible explanation for the presence and form of the solar system. It accounts adequately for the fact that the planets share a common plane, for the uniform direction of rotation and revolution of both planets and satellites, and for the loss of angular momentum. The relatively slow rotation of the sun remains unexplained, yet it may be possible to explain this within the general framework of the theory.

The process that the theory sets forth is regarded as representing a normal stage in the life cycle of most stars, and it remains valid when applied to binary and multiple star systems. In this respect Gerard Kuiper has shown that the gas could condense into two or more nuclei that begin to revolve about one another. At other times, of course, the gas merely develops into a single sun. The remaining scraps of matter, being too small to support the thermonuclear reactions of a star, become planets. Fragments such as these would vary greatly in size, and indeed their dimensional range could be considerably greater than that which we find in the solar system. The star 70 Opiuchi, for example, is believed to possess at least one planet considerably larger and more massive than Jupiter.

This process should not therefore be regarded as unique. It happened to a star we call the sun, it has happened to other stars, and to some stars it is probably still happening. Therefore, if we look around the universe it seems reasonable to expect some confirmation. There is of course a fundamental and practical difficulty. The stars are so incredibly remote that no matter how large attendant planets may be we cannot hope to see them. Even the great 200-inch telescope at the Mt Palomar Observatory cannot help us here. Perhaps in the future when comparable (or greater!) instruments can be erected on the airless surface of the moon it may become possible to discern such worlds. However, the presence of satellite bodies (i.e., planets) in the close proximity of other stars has already been confirmed in a few instances. The existence of these is revealed by slight but significant periodic variations in the predicted positions of the stars concerned. Such variations are attributed to the gravitational pull exerted by smaller, invisible, and much less massive companions. For a number of years the two classic examples were the stars 61 Cygni and 70 Opiuchi. In 1963 Peter Van de Kamp in the United States was able to confirm planetary presence around Barnard's Star.

The detected planets of these stars are very large and massive. Were this not so, it is most unlikely that their existence would have been revealed. It is virtually certain that worlds comparable in size to the planets Earth, Mars, and Venus could not be detected in this way. The pull exerted on parent stars by planets of this size is simply too negligible to be apparent. Where large planets exist, however, so also can much smaller ones. For proof of this we have only to consider the relationship between Jupiter and Earth, the former being 318 times more massive. Su-Shu Huang maintains that the masses of bodies created from a primordial nebula must range from those of stellar dimensions to small bodies comparable to Earth and Mercury.

That there may be a direct connection between the rotational speed of stars and the existence of planetary systems seems possible. Very massive hot stars (notably those in spectral classes O and B) are rotating with velocities of the order of 100 to 300 kilometres per second. Class A stars, which are both cooler and less massive, show rotation velocities in the range of 40 to 90 kilometres per second. As we continue along the spectral scale this trend becomes increasingly apparent, as Table 1 shows.

Spectral Class	Rate of Rotation (kms/sec.)	Proportion of Galactic Population (%)	
O	100–300	0·35	
B	100–200	11·00	
A	40–90	22·00	
F		18·00	
G		14·00	
K	0–10	31·00	66·75
M			
R, N, S		3·75	

TABLE 1

Thus main-sequence stars of classes G, K, and M, whose temperatures are below 7,000 degrees centigrade and whose masses are less than 1·2 of the sun's mass, possess equatorial rotation speeds that are quite low. As an added point of interest it should be mentioned that our own sun is a main-sequence star of class G.

Although mass, brightness, and surface temperature show a fairly regular gradation, there is a very sharp dip in the equatorial rotation rate. This dip, as the table shows, occurs between class A and class F. It is now considered possible that the unusually low rotation rate among main-sequence stars of these latter classes is due to the motion around them of non-visible bodies of small mass (i.e., planets) that have orbital rotation rates several times greater than the axial rotation rate of the parent star.

Let us apply this for the sake of argument to our own sun, whose equatorial rotation rate is 10 kilometres per second. Were the total momentum of the solar system to be concentrated in the sun alone, the equatorial rotation of that star would be in the region of 100 kilometres per second, which is more or less akin to that of a class A or class B star. Thus there is a rather obvious inference. Stars of classes F to S possess slow rotational rates *because of the existence of planetary systems.* Indirect observational evidence points to the probability of planetary systems in a number of cases, and this is reinforced by the foregoing theoretical reasoning.

It seems therefore extremely probable that a high proportion of main-sequence stars from class G onwards (i.e., K, M, R, etc.) will be accompanied on their journeys through space by planetary retinues. This being so, the number of planetary systems within our galaxy must be of the order of several billion, and that of individual planets even greater. It is possible to go still further. If our sun is regarded as the centre of an imaginary sphere having a radius of 100 light years, this particular pocket of space will perhaps contain something of the order of 10,000 stars. Of these stars a high proportion are main-sequence stars of the spectral groups already specified. This is a factor of tremendous importance and implication.

The existence of planets does not automatically mean the existence of life, intelligent or otherwise. Nine known planets orbit the star we call the sun yet it is fairly certain that at the present time only one, our own familiar Earth, contains living beings of an advanced intelligent form. There may be vegetation of a low order on Mars, and it is even very remotely possible that some form of life may exist on Venus. Nevertheless, to the best of our knowledge at the present time only one planet of the nine contains intelligent life.

Is this, then, typical of solar systems throughout the universe? To that question it is impossible to give a definite answer, for truthfully we do not know. There is, however, a case for believing that in its general characteristics our own solar family may be fairly representative. Life requires a rather narrow set of essential conditions. Planets must be neither too hot nor too cold. They must possess suitable atmospheres. They must have a rate of axial rotation sufficient to ensure a fairly even distribution of heat and cold. By reason of gravitational attraction planets must be neither too large nor too small. There must also be a reasonable amount of dry land on which life may take root, develop, and exist. The last may not be an essential prerequisite for marine life, but it seems that advanced civilizations and technologies of an ichthyological nature are on the whole less likely!

This concern, of course, leads to another important aspect. Must all advanced forms of life be moulded on the pattern of our own? The answer within certain limits is no. Nevertheless we must always take care not to give this particular argument unqualified licence. Planets of other suns will in many cases possess characteristics quite different from our own. Life will, therefore, be shaped by the environment into forms best suited to prevail. At many points within the galaxy this could lead to alien creatures weirdly different from ourselves.

Given a planet of suitable conditions, we remain far from certain just how life originates. Indeed we do not even know how it was initiated on our own world, though recent biochemical research has shed some light on the subject. We can safely assume, however, that the formation of the necessary complex organic molecules must have occurred at a very distant period in the history of this planet. Through millions of years these organisms gradually evolved. *Homo sapiens*, our own kind, appeared on the terrestrial scene at a comparatively late stage in our world's biological evolution.

Because of the time scale involved it is easy to see how the universe must almost certainly contain life at varying stages of development. We must anticipate planets where life has either not yet begun or is just beginning. There may well be worlds out among the stars where terrifying monsters, the equivalent of our prehistoric giant lizards, roam desolate wastes under the strange eerie light of alien suns. We can envisage other worlds where 'Stone Age' men dominate the scene and others where reasoning creatures have reached a point in mental development akin to our own. The most fascinating prospect of all is that of worlds whose peoples are further advanced than those on Earth – peoples who have raised civilizations and created technologies beyond our wildest imaginings. At both ends of this biocosmic time scale we can and indeed must extrapolate.

This planet of ours is believed to be some 5,000 million years old, although life has existed on it for only part of that time. Therefore it is possible for worlds whose physical characteristics are appropriate to be still devoid of life. What happens at the other end of the scale? Does life eventually peter out? If surface conditions deteriorate sufficiently this certainly would seem inevitable. An obvious example would be the case of a planet whose mass is insufficient to permit the permanent retention of a suitable atmosphere and whose essential gases gradually leak away into space. It may be that Mars falls within this category, for we know that the Martian atmosphere is by now very attenuated indeed. Is it possible that the red planet was once inhabited by the race of master engineers and dedicated scientists so beloved of the science-fiction writer?

Let us, however, contemplate worlds that have managed to retain sufficient atmosphere. Does life die out for other reasons or undergo such a measure of evolutionary change as to render it totally different? Reluctantly we are forced to admit that we do not know the criteria for the continuing existence of living beings and their civilizations. If life on a planet eminently suited in all respects does eventually disappear, we are confronted with the almost certain existence of worlds devoid of life for the very simple reason that there life has run its allotted course.

An interesting analogy has been worked out by a British amateur astronomer, Patrick Moore, to illustrate this point. Consider for a few moments a darkened room containing two electric lights. Both of these lights are switched on for only 10 seconds in every 24 hours. Moreover, these 10-second periods are selected wholly at random for each light. Clearly the chance of both being on simultaneously is extremely slight. Nevertheless, this is said to represent a greater possibility than that of two civilizations existing in *adjoining* stellar systems at the same time!

Such a fact must have a distinct bearing on electronic communication between 'close' or neighbouring star systems. At first it may seem more than a little discouraging, but we must remember that the stars are many and as we go towards the hub of the Milky Way their numbers increase to an almost unbelievable extent. This, of course, involves infinitely greater distances. Signals from such remote sources will be very much weaker and will also take correspondingly longer periods to reach us. Nevertheless, the number of stars 'close' to us in the cosmic sense is still sufficient for us to retain a moderate but justified measure of optimism.

There is another important point that must be mentioned at this stage. Earlier we enumerated certain essential physical conditions that a planet must maintain in order to spawn and nurture life. We spoke of distance from the central luminary, the presence of a suitable atmosphere, and the right measure of both gravity and axial rotation. We must, however, recognize the fact that many stars are vastly different from the sun in their physical characteristics. We are thinking not of stars whose type or age render them incapable of promoting life but of stars whose physical states impose a greater impediment to life on planets in their possession. This category includes stars that have an extreme variability and also certain types of binary and multiple systems. Both of these groups introduce the possibility of sharp and probably catastrophic temperature variations.

We are now able to provide a more specific summary, as follows:

1. There is a firm basis for assuming that throughout the galaxy billions of planetary systems are scattered and that in at least a proportion of systems life will have evolved.

2. Such planets must lie at a suitable distance from their respective suns, and their surface temperature must fall within the range necessary for biological initiation and development. Since a number of planets must originate with or be born from a star, it is reasonable to anticipate that at least one or two planets in a large number of planetary systems will have temperature ranges of the appropriate order.

3. The mass of a planet must be neither too great nor too small. Very large planets will very probably possess hydrogen-

rich primeval atmospheres in which it is unlikely that life can prevail. We cannot, of course, state with absolute certainty that forms of life suited to such atmospheres are completely impossible. Such planets, however, will have high gravities, a feature hardly compatible with the concept of active, developing life. In the case of very small planets with low gravities such atmospheres in the fullness of time may disperse instead of undergoing transition to the oxygen–nitrogen type of atmosphere on which we rely on Earth. Small planets having suitable atmospheres are, however, much more likely to be closer to the central star than large, massive planets with atmospheres composed of hydrogen and hydrogen-compounds. Consequently, temperature suitability is likely to be largely 'in phase' with atmospheric suitability.

4. Advanced life and the complex civilizations that go with it are most likely to be found on planets orbiting 'old' stars, that is, stars whose age is of the order of many billions of years. The reason for this is that life takes a very long time to evolve and develop. Fortunately, the stars in the spectral groups that we have already specified satisfy this essential condition. (See Table 2.)

Spectral Class	*Age (Years)*
O	$10^5 – 10^6$
B	$10^6 – 10^7$
A	$10^7 – 10^9$
F, G, K	$10^9 – 10^{10}$
M, R, N, S	10^{10}

TABLE 2

Earth is thought to have been sterile for some 4×10^9 years (4,000 million) after its formation. It is therefore very probable, even allowing for favourable conditions, that life will not appear on any planet that has been in existence for less than a minimum period of $(1–4) \times 10^9$ years. Cosmology indicates that the age of a planet must be akin to that of the parent star and directly related to its spectral category. Thus, as Table 2 shows, stars in group F to group S (those most likely to possess planetary systems) come within the necessary 'life' period.

5. The star concerned should in each case be of a stable nature over a period of several billion years. This condition is satisfied by the greater proportion of stars of interest to us.

6. It is preferable that the particular star be either a single entity or else a member of a binary or multiple system whose individual components are well separated. If the distance between components is sufficiently large, planets can be expected to orbit only one of these components much as the planets of a single star system do.

Since there would appear to be a fairly good *prima facie* case for the existence of life-bearing planets, the next step is to assess, if possible, the proportion of those other worlds that might be expected to contain life. A real assessment is, of course, impossible, the most we can hope to attain being a figure based on a reasoned guess. (See Table 3.)

Number of stars in our galaxy (the Milky Way)	$(1-2) \times 10^{11}$
Estimated number having planets	10^{11}
Number of planets (based on solar system total)	10^{12}
Number of planets at suitable distance from star	10^{10}
Number of planets large enough to retain an atmosphere	10^{9}
Number of planets having a suitable atmosphere	10^{8}

TABLE 3

Thus within our galaxy there are probably 100 million planets capable of nurturing a form of life. This estimate is a fairly conservative one. Others have been made that place the figure much higher, and one estimate actually puts it at 1,000 million.

We have already dwelt on the idea of life having a limited existence on any planet. Certainly it is difficult to conceive of terrestrial man continuing indefinitely in his present form. Intelligent life might prevail on a planet for millions of years, or its tenure could well be of an appreciably shorter duration. An entirely arbitrary figure of *one million* years would imply that at the present time something of the order of several million planets could possess life of a highly advanced kind. This in turn would mean that within a radius of 100 light years of the solar system there might only be a relative handful of life-bearing planetary systems.

It is possible, then, that perhaps ten million planets within our galaxy contain life at the present time. Since the majority are too remote to be of more than academic interest, it is necessary for us to restrict ourselves to the aforementioned imaginary sphere that has a radius of 100 light years. Although this drastically reduces the number of possible life-bearing systems with which contact may be sought, it should not be regarded too adversely because we can feel fairly certain that the handful of alien civilizations surrounding us are nevertheless very real.

One of the most fundamental aspects of this problem is the relationship between the periods necessary for stellar and biological evolution respectively. We must start by considering the time required for an advanced and intelligent life form to evolve. In the case of Earth, this was of the order of many millions of years.

The time scale of stellar evolution is a very different matter indeed. We saw in Chapter 3 the manner in which a star is formed, and dwelt briefly on the processes involved during its amazing life cycle. We recall that a normal star spends the greater portion of its life in what is termed the main sequence. It is during this period that a star is most likely to possess life-bearing planets. Prior to that period planets are too young for life to have appeared. Afterwards, the career of a star becomes altogether too hectic for life to continue and even perhaps for the planet itself to continue. The time, therefore, in which a star remains in the main sequence must be very much greater than the time required for biological evolution. If this condition is not satisfied, any life that has made its appearance on a planet of that star will be completely eradicated by the normal evolutionary process of the star long before this life has had a chance to develop.

The time spent by the various star groups in their main-sequence stage varies considerably. In the case of group O stars it is of the order of 10 million years, but by the time class M is reached this has increased to 100 thousand million. Perhaps at this point it would not be out of place to enumerate again the various groups of stars. They are:

WOBAFGKMRNS

(For those who find difficulty in remembering this unusual order of letters the following, now almost immortal, mnemonic is recommended: *W*ow, *O*h *B*e *A* *F*ine *G*irl, *K*iss *M*e *R*ight *N*ow *S*weetest!)

We saw earlier that planets are most likely to be found orbiting stars of class F onwards. We can therefore eliminate stars of class O, which have a 'short' main-sequence career of only 10 million years. Class A and class B can also be forgotten because these too have relatively brief main sequences. It is class F to class M that are really of interest, for in this instance main sequences endure from 10,000 million to 100,000 million years. We find therefore that our previously specified essential condition is well fulfilled since biological evolution involves a considerably shorter period of time than this. In the case of Earth a recent estimate puts it at 1,000 to 3,000 million years.

A weakness, or apparent weakness, in this argument may seem to lie in the fact that we are using Earth as a standard. Unfortunately terrestrial life is the only life we know, and there is no valid alternative. Biologists tell us that evolution, a result of mutations, is a random and therefore slow process. Consequently the case of our planet may be fairly typical. Of course it could be asked whether or not other stars, because of more intense radiation, might not accelerate the process. This is another question we are unable to answer even though present biological evidence runs contrary to such a belief. The consensus seems to be that the process of life initiation and development will proceed on any suitable planet at a rate not greatly dissimilar to that of our own process.

The next point concerning life is almost as important, and we certainly need to go beyond a study of our own familiar solar system in this instance to derive the essential truths. Our sun possesses, so far as we can observe at present, nine major planets. Here the term 'major' is used in a broad context to differentiate between these nine worlds and the minor planets and planetoids or asteroids whose life-bearing potential must almost certainly be zero.

Let us divide the solar system into three zones. The first zone includes the sun itself and extends almost to the planet Venus. This zone we must regard as being simply too hot for carbon-based life to prevail. The second zone takes us from the orbit of Venus to just beyond that of Mars. This is the zone where life is possible and where Earth lies more or less in its centre. The third zone takes in the orbit of the giant planet Jupiter and extends outwards to Pluto. Within this zone temperatures are much too low for life to be even remotely feasible. Those science-fiction writers who in the past have enthralled and entertained us with tales of Jovians, Saturnians, and even Plutonians must realize that they have conjured up very remarkable living beings indeed!

Today we believe that the solar system is fairly typical of such systems. No doubt there are stars whose planets number less than nine and many also that have more than nine. Each system will however have its 'habitable' zone. Life could not be spread uniformly or indiscriminately through other solar systems any more than it is through our own.

In the case of stars of a type and size similar to the sun their life zone will be comparable in extent to the one in which we on Earth exist and thrive. Such systems might not, of course, have a planet or planets conveniently placed in the middle of this zone, and in these cases the question of life would be very doubtful. This, however, is a speculative point, and it seems on the whole unlikely that such a wide zone would be completely devoid of planets.

Where stars markedly different from our own are concerned, the position is considerably more fluid. Were a much hotter star than our sun to lie in the centre of the solar system the zone that we regard as habitable would be, to say the least, most inhospitable. Under these conditions this zone of excessive heat would extend much further than the orbit of Venus, and the region of Saturn to Pluto might well constitute the habitable one. A small, 'cool' star in the place of the sun would, on the other hand, probably render Mercury potentially habitable but banish Mars and perhaps even Earth to the region of eternal, frigid cold.

Using the solar system as an example might give the erroneous impression that these zones, habitable or otherwise, are in the form of concentric rings at the centre of which lies the sun.

These so-called zones are, however, in reality concentric shells or spheres. The fact that all our planets lie in the same plane does not mean the worlds orbiting the sun in a plane at right angles to the existing one would not also occupy one or other of the particular zones we have discussed.

The more luminous a star the greater should be its habitable zone and, therefore, its ability to sustain life. Thus stars in the lower portion of the main sequence will have less capacity in this respect. This applies particularly to those in the late K and M grouping. These are fairly obvious facts to take into consideration when we come to survey the heavens for suitable stars towards which we can direct our radio-telescopes. Faint stars are clearly not a good bet in this respect. By faint, of course, is meant stars that are intrinsically dim, not stars faint to our eyes merely by reason of distance. This does not necessarily mean that dim stars will not have planets, but any such planets will be considerably less likely to have spawned life.

We have tended so far to regard planetary orbits as being invariably circular. This need not always be so. Some orbits within the solar system are elliptical, and there are sound reasons for regarding this as a perfectly normal state of affairs. An elliptical orbit means that there will be times when a planet is closer to its parent star than at other times. It is therefore possible under certain circumstances for a planet to lie within a habitable zone for part of its orbit and outside the zone for the remainder of its orbit. This would almost certainly invalidate that planet as a potential centre for life.

An exception to what has been said regarding intrinsically faint stars might possibly be provided by red dwarf stars. The habitable regions of such stars could hardly be extensive, but since red dwarfs are known to remain in the main sequence for a very long period we might reasonably expect living creatures to have evolved in some instances.

A high proportion of stars, especially in the immediate environs of the sun, are known to be members of binary or multiple systems. To what extent would this influence life on the planets of such stars? The idea of worlds in which two or more suns, probably of contrasting colours, appear simultaneously in the sky is one that has intrigued both readers and writers of science fiction for a long time.

It is generally believed, however, that binary and multiple systems tend on the whole to be inimical to life. Planets orbiting double or multiple star systems are extremely likely to have eccentric orbits. That such worlds might wander in and out of habitable zones seems a reasonable premise in these circumstances.

This might be less true in the case of a planet orbiting only one component of a binary or multiple system. Much, of course, would depend on the distance separating the individual stars of the system. It must be conceded, however, that though a planet might orbit merely one of the component suns, the orbit of that planet could still be seriously disturbed by the other sun. Consequently such a planet might for part of its course be drawn into a non-habitable zone.

The concept would seem to be that in the case of binary and multiple star-systems the prospects of life *are* materially reduced. Nevertheless, in view of the profusion of such systems, the possibility that some may possess life-bearing planets cannot be wholly discounted. In the case of a close binary system we would expect life-bearing worlds to be exterior planets since their orbits would thereby tend to be less disturbed. Were they, on the other hand, to lie too far out, their role might be that of permanent residents of a non-habitable zone. Distant binaries (i.e., those whose individual components lie far apart) would probably have their life-bearing planets set reasonably close to one or other component.

Since from a communication point of view it is the nearer stars that most interest us, it may be profitable to examine this aspect more closely. If we regard the sun as the centre of an imaginary sphere we find that within a radius of $16\frac{1}{2}$ light years there are 59 other stars. This total number includes 24 single stars, 13 double stars, and 3 triple stars. It is therefore more accurate to say that 40 other *stellar systems* lie within $16\frac{1}{2}$ light years of the sun. Some of these binaries and multiples include white dwarfs. In addition, three of the stars have an extremely high velocity and should, therefore, be looked upon as mere temporary members of our little stellar colony since, of course, they will eventually leave it. One of these is 61 Cygni, already thought to possess at least one planetary companion. In the relatively short space of 400 years this star has moved a distance apparently equivalent to the diameter of the moon. Another is the well-known Barnard's Star whose velocity is greater still – it is a real hustler even by celestial standards. This star also displays evidence of having a planetary companion. Presence of this remote and mysterious world, about which as yet we know virtually nothing, was finally confirmed by Van de Kamp as recently as 1963.

The two stars believed most likely to possess life-bearing planets are Epsilon Eridani (10·8 light years distant) and Tau Ceti (11·8 light years). Both stars have less than half the brilliance of the sun and belong to spectral groups K and G respectively. Both are single stars.

It is hardly surprising, therefore, that it was these stars that were chosen for examination in Project Ozma, the first systematic search for intelligent alien beings ever instituted on this planet. This work was carried out during May, June, and July 1960 by Dr Frank Drake and his colleagues. In their project experiment they used the 85-foot radio-telescope of the National Radio Astronomy Observatory at Green Bank, West Virginia. Whatever the future may bring, there can be little doubt that the year 1960 will remain a milestone in the technological and cultural history of the world. For the very first time man cast aside some of his more deep-rooted prejudices and looked out on the eternal, limitless cosmos with an open mind. It is to be hoped that this represents the true spirit of the future, for if it does that future is full of hope.

No signals attributable to extraterrestrial, extrasolar origins were detected, which had been largely anticipated. This was after all merely a preliminary sally, the harbinger of greater things to come.

It is perhaps natural at this point to think of Alpha Centauri, the closest of all stellar systems. To those of us who live permanently in the Northern Hemisphere this star is but a name. It is, however, one of the brightest stars in the terrestrial heavens, and the writer can well recall its brilliance seen under a South African sky in 1943. Alpha Centauri is actually a triple star system, the two most massive components of which are G class and K class stars revolving around a common centre of gravity. The third component, the renowned Proxima Centauri (actually the closest of all stellar bodies to the sun), is an M class dwarf.

The eccentricity of the binary orbit is quite considerable, and for that reason it is preferable to omit this system for the present. There may well be planets there, even habitable ones, but both Epsilon Eridani and Tau Ceti are more promising in this respect despite their greater remoteness.

The next important point is the relationship between the size of a planet and its degree of habitability. Once more we are of necessity compelled to use our own world as a standard since it is the only planet of which we have intimate knowledge.

Biologists believe that a liquid medium is essential in the first instance for the production of living cells. It is clear, therefore, that a suitable planet will be one having a solid crust and the ability to retain water.

Planets evolve from the same material as their parent star does. In evolving they pass through a transition stage in which they contain a very high percentage of hydrogen in their outer envelopes. Though life cannot be regarded as impossible in a hydrogen atmosphere, we must admit that the likelihood is considerably more remote, especially in respect to higher life forms. Atmospheres of hydrogen and hydrogen compounds are regarded as being primary or transitional, being replaced eventually by those of a secondary nature. This is believed to have occurred countless millions of years ago in the case of both Earth and Mars. When we look towards the giant planets of the solar system – Jupiter, Saturn, Uranus, and Neptune – we see worlds whose atmospheres are presently in this primary category. Jupiter's gaseous envelope contains hydrogen, methane (CH_4), and ammonia (NH_3), as does Saturn's envelope. The atmospheres of Uranus and Neptune are composed predominantly of hydrogen and methane. It may well be that after the passing of countless aeons these atmospheres will undergo transition, too.

A prime requisite in the case of habitable planets is that their primary atmospheres be transformed into those of that secondary type in a period that is short compared with that of stellar evolution. If this does not occur, the time for an evolving life runs out, since that life would begin too late to avoid extinction when the star swells out during its post-main-sequence stages. This period, it is believed, varies from 10^9 to 10^{11} years (1,000 million to 100,000 million). This permits an upper limit to be placed on the mass of a habitable planet. Planets whose mass is so great that they retain a primary atmosphere for a period in excess of 10^9 to 10^{11} years must be regarded as unsuitable. Our own Jupiter and Saturn must be placed in this category. However, since these worlds lie very far from the sun, their inherent low-surface temperatures already preclude any real life potential.

Secondary atmospheres must also be regarded as transient. It is, therefore, equally essential that these atmospheres should not have left the scene before living organisms appear upon it. The time required for the evolution of terrestrial life is thought to have been $(1-3) \times 10^9$ years (1,000 million to 3,000 million).

It has been suggested that to be potentially habitable a planet should have a diameter between 800 and 16,000 miles. This, however, is almost certainly a much too generous estimate so far as the lower limit is concerned. The upper limit of 16,000 miles, representing as it does twice the diameter of our own planet, seems on the whole not unreasonable.

It would be possible to dwell at considerable length on the many factors concerning extrasolar planets, but our thesis in this book is primarily that of radio contact with star peoples. What has been said in respect to stellar planets should, however, be sufficient to spotlight the inherent possibilities and therefore to stress the plausibility of ultimate interstellar communication.

It is inevitable that much speculation must attend the subject of the Earth's cousin worlds. Nevertheless, it is now almost universally conceded that such worlds must exist and that in some cases they will be of a type potentially suited to life on the terrestrial pattern. The possibilities of life, even supra-intelligent life along non-terrestrial lines, must also be envisaged. This does not necessarily endorse the imaginings found in the more lurid science fiction. At the same time a reasonably open mind must be kept. It may well be that if extensive electronic contact is ever made with alien peoples the truth of an old adage will be further stressed – that truth is indeed stranger than fiction!

5 Creatures of the Stars

THE origin of life is a specialized and highly involved subject about which only biologists are fully competent to speak. Its implications are far-reaching, and an entire work could easily be devoted to it. A full treatment of the subject is beyond the scope of this book, and our aim can only be to dwell briefly on its broad essentials.

So far as the form of intelligent beings is concerned, it might be said that there are in existence two distinct schools of thought. One school claims that, by and large, the human form is ideally suited to our environment, and that all planets remotely like Earth should be peopled by erect, thinking bipeds having two arms, two eyes, two ears, and so on. The other school, however, suggests that this is really the height of conceit and that form, even on terrestrial-type planets, must vary considerably.

It is, to say the least, most difficult to decide which of these conflicting opinions is the more valid. Certainly it is hard, so far as terrestrial-type planets are concerned, to conceive of a shape and form more suited to the prevailing conditions than our own. We are air-breathing creatures whose essential organs are protected behind a framework of bones. We have been given two legs that are well suited as a means of locomotion over rough or uneven terrain. Our arms are ideally placed, and our hands and fingers enable us to grip and handle all kinds of solid objects. The vital brain is positioned at the top of the body where it is less likely to suffer damage. The brain is also protected fairly adequately by the skull. Also in this region, in close and convenient proximity to the brain, are located the organs of sight, speech, and hearing.

While this may be a brief and very sketchy description of the human body, it does serve to show that its form has been largely dictated by essential requirements and is not merely the result of nature's whims. We must remember that our form evolved. Through the later aeons of the Earth's existence, life has been doing just that – evolving, rendering itself appropriate to its environment. Life, we believe, started in the primeval oceans; eventually some form crawled ashore, some fish became amphibians, and these amphibians ultimately became creatures of the dry land. In the end (if it is the end) came man, not the supreme terrestrial being by virtue of physical strength but rendered so by the possession of a brain that has a considerably greater capacity than that of any other living species. During this long and somewhat eventful biological journey, nature made mistakes by producing forms unsuited to the planetary environment. There is, for example, the instance of the giant reptiles, which were ruthlessly and abruptly eliminated from the scene.

Nature has adapted life to survive, to thrive, and to continue under the conditions prevailing on this planet. What of life on worlds where conditions are quite different from those on our own planet? If these are too different we believe that life would not be possible and that such worlds must remain forever sterile. What, however, of those planets where it is possible for life to prevail in conditions markedly different from those of Earth? Here life will surely follow distinct paths of its own, adapting itself by degrees to the environment in which it finds itself. Intelligent creatures on such worlds, though they might have much in common with ourselves, would certainly not be identical in outward form. The degree of difference could indeed be considerable.

There is so much sheer speculation involved in philosophy of this sort that any kind of conclusion is impossible. On worlds closely akin to our own, life if it exists may possess forms very similar to those on Earth. On worlds of a quite different type, where advanced life is nevertheless possible, evolution may have followed a very different course. What we are saying in effect is that both schools of thought are to some extent correct. Life on planets like Earth *may* be akin to what we are accustomed to, but *all* life throughout the galaxy will not be moulded on such lines.

No one dares to categorize precisely the manner in which life was initiated on this world of ours. We can, however, have a brief look at a few ideas that seem to be not unreasonable. In a number of cases we can even point to a degree of substantiation by practical experiment.

The kernel of our problem can be summed up in one word – biopoesis, which is the process whereby living matter can be produced from non-living matter. We must try to ascertain whether the conditions prevailing on that strangely alien and primeval Earth of 3,000 million years ago could have touched off the chain of events that led to the teeming life of today. We might ask whether life was initiated by some happy chance combination of chemicals or whether it must be regarded as the result of a general scheme repeated over and over again throughout our galaxy and all the countless others. Writing in his book *Origin and Nature of Life* (1913), Dr Benjamin Moore, Professor of Biochemistry at the University of Liverpool, said:

> It was no fortuitous combination of chances, and no cosmic dust which brought life to the womb of our ancient mother Earth in the far distant Palaeozoic ages but a well-regulated orderly development which comes to every mother Earth in the Universe in the maturity of her creation, when the conditions arrive within the suitable limits.

Today, although we do not claim to understand life-initiation processes or the exact conditions necessary for them, we do endorse much more emphatically the idea that life is not just a cosmic mistake, not merely the result of some impurity or catalyst in nature's elemental crucible.

The first point to consider is that of life initiation, and once more we are of necessity forced to use our own planet as a standard. In the 1920s we find both J. B. S. Haldane and A. I. Oparin postulating at some length on terrestrial life's beginnings. The former held the opinion that Earth's primary atmosphere contained little or no oxygen, a view that is generally accepted today. He also maintained that the early atmosphere had been extremely rich in carbon dioxide (as is that of Venus at the present time) and that a high proportion of the free nitrogen presently found in our planet's atmosphere had existed within Earth's crust in combined form (probably as nitrides). He further maintained that owing to the absence at that time of oxygen (or of its allotrope, ozone, O_3) little of the sun's intense ultraviolet radiation was filtered out.

Haldane may well have been influenced by the results of certain experiments that had been carried out shortly before he formulated his theory. These experiments had demonstrated fairly conclusively that the effect of subjecting a mixture of water, carbon dioxide, and ammonia to ultraviolet radiation is the production of a number of organic compounds including, rather significantly, a few simple amino acids. It is easy to see in this experiment the analogy of nature at work on the grand scale. Our laboratory vessel of water, carbon dioxide, and ammonia becomes violent, primeval earth with its turbulent atmosphere of carbon dioxide and ammonia, its roaring oceans already transformed into dilute solutions of the former compounds. Our source of ultraviolet light is now a much younger and more virile sun transforming the early seas of our world into great broths of 'life-potential' chemicals.

Oparin's ideas on the subject did not really differ significantly from those of Haldane except perhaps on one issue. Haldane had believed that carbon in the primary atmosphere was present as carbon dioxide (CO_2), whereas Oparin remained firmly convinced that methane (CH_4) was the more likely gas. Some substantiation for Oparin's view is afforded by the fact that the giant planets of the solar system, Jupiter, Saturn, Uranus, and Neptune, all have atmospheres containing a high proportion of methane.

On one point we can be fairly certain. Oxygen, were it present at all, could only have been so in relatively minute amounts or it would quickly have combined with appropriate materials in the earth's outer crust. Indeed its presence in our atmosphere today is remarkable considering its chemical affinity for other elements.

Haldane and Oparin had both concentrated on ultraviolet radiation from the sun as the agent transforming inorganic materials into organic. S. L. Miller, however, later demonstrated that lightning and corona discharges were also able to produce transitions of this nature. The original gas mixture in Miller's experiments was similar to the mixture Haldane employed except that carbon dioxide was now absent, having been replaced by methane (as in Oparin's experiment). Hydrogen was also added. Once again several organic compounds were produced, among them a number of amino acids. Most significantly these amino acids were *not* produced on occasions when free oxygen was added to the original mixture.

Thus far, then, we see how both ultraviolet radiation and electrical discharges will convert appropriate inorganic mixtures into solutions of organic compounds – compounds known to have a considerable bearing in the field of biochemistry.

A third distinct agency is known to produce a similar effect – bombardment by subatomic particles, that is, alpha, beta, and gamma radiation. The accompanying table briefly summarizes the position in this respect.

Radiation	Reactants	Products	Details
Alpha	carbon dioxide, carbon monoxide, and hydrogen	resinous organic materials	Lind and Barwell 1962
Alpha	carbon dioxide, carbon monoxide, and methane	resinous organic materials	Lind and Barwell 1962
Alpha	carbon dioxide and water	formic acid and formaldehyde	Garrison
Beta	ammonium acetate	glycene and aspartic acid (both amino acids)	Hasselstrom
Gamma	ammonium carbonate	glycene (among others)	Paschke

TABLE 4

Alpha and beta radiation, so far as the primeval Earth is concerned, could easily have been provided by radioactive minerals present within the Earth's crust.

In the face of this experimental evidence, it seems most probable that alpha, beta, and gamma radiation, cosmic rays, and electrical discharges led to reactions among inorganic materials and simple hydrocarbons (e.g., methane, CH_4) thereby producing organic compounds known to be essential to the development of life.

Further experiments by A. P. Vinogradov brought up once again the earlier differences between the views of Haldane and of Oparin. The former, it will be remembered, adhered to the idea of carbon dioxide in the primitive atmosphere, whereas the latter opted for methane. Vinogradov, however, suggested that the gaseous carbon compound involved was in fact carbon monoxide (CO). This conclusion was reached because an analysis of the gases emitted by volcanoes showed the presence of carbon dioxide and carbon monoxide but not methane. On the whole, this view appears to have found little acceptance.

Whether the carbon compound concerned was carbon monoxide, carbon dioxide, or methane does not matter very much since the agencies already specified appear able to transform atmospheres containing these gases into more complex and vital organic compounds as long as free oxygen remains absent. Therefore amino acids could have been created from the primitive atmosphere by agencies that we can be sure did exist at that time.

The next link in the chain of life is a group of compounds known as proteins. These compounds are considerably more complex in their structure than amino acids, which can be regarded rather loosely as the essential raw material for the manufacture of proteins. Much can be said about proteins, but it is hardly necessary to go into such detail here except to remark that they are an essential component of living organisms. Thus it now becomes possible to appreciate the basic significance of amino acids and all that has so far been said regarding their creation.

More than a score of amino acids have been positively identified in proteins, but the two principal ones are undoubtedly aspartic acid and glutamic acid (25–50 per cent of the total protein). It is further known that these two acids are of prime importance in biological processes. The validity of this can be demonstrated in the laboratory by heating a mixture of amino acids in which glutamic and aspartic acids predominate. If the temperature is maintained at around 180 degrees centigrade for approximately three hours a definite protein-like substance results.

Much interesting experimental work has been done along these lines, and this is clearly a line of research that may one day yield startling results. Whether or not it will ever become possible to produce living cells artificially no one can yet tell. With increasing knowledge this may become possible. Inevitably this raises many important issues including those of a moral and religious nature, and much thought must certainly be given to such aspects. It does, however, seem unlikely that any really complex living organisms could be created artificially.

We should not be unmindful of the important though secondary roles that other naturally occurring materials may have played in this strange drama. There are, for example, the possible catalytic effects that could conceivably have been brought about by quartz (silicon dioxide, SiO_2) and aluminium silicate. It is by no means unlikely that these materials and others of a clay-like nature present in the primitive oceans of Earth played a vital role by adsorbing particles of potentially reactable chemicals. In so doing they may well have brought molecules into much more intimate contact. This would certainly contribute to the establishment of optimum conditions for the reactions. Other minerals, too, might have assisted in these reactions by a purely catalytic effect, that is, without adsorption.

Early rain has also been suggested as a possible initiating agent, though, of course, the rain referred to is not the relatively gentle variety of today. This early rain was of a torrential kind and may have contained quantities of oxygen and peroxides absorbed from the upper atmosphere. That it could have promoted or triggered off the necessary reactions is presumably possible though this must be regarded as rather speculative.

The biology or biochemistry we have just been discussing relates specifically to one planet, our own. It should be understood that if these processes, or ones akin to them, did take place on Earth, then they should also have occurred (or should be occurring) on planets where the prevailing circumstances are favourable. Before we look more closely at this aspect, however, it is advisable to consider further the subject of life on our own planet.

The gradual climb of terrestrial life up the evolutionary ladder was necessarily slow. The first scraps of living organism almost certainly must have been as frail and helpless as blobs of jelly. The first distinct organs to evolve were probably those necessary for absorbing nourishment, discharging waste, and for moving about.

In the early ages, life on Earth was entirely restricted to the oceans. After the passage of some 300 to 400 million years it is believed that the climate became markedly drier. So far as the fish that had restricted themselves to the great oceans were concerned, the fact that the climate was drier was of little or no significance. Many, however, had already migrated to rivers and into fresh-water lakes, and for these fish the climatic change was of considerable importance. The water in small ponds and lakes gradually evaporated causing dire consequences for their occupants. In many instances certain of these creatures had already adapted themselves to crawling short distances over dry land and were thus able to take themselves to bigger, deeper pools that had not dried up.

Eventually even the deepest and biggest of these fresh-water pools began to dry up, which compelled the occupants either to re-adapt or to perish. To re-adapt, of course, meant acquiring the ability to live permanently on dry land. The creatures that did re-adapt in a broad sense represent our common ancestors.

The process of adapting from an aqueous medium to dry land could not have been easy. For some, the organisms that had already learned to live ashore for brief periods, the transition was probably more straightforward. For all, however, the new environment was basically hostile, and evolution sought to counter this by surrounding these creatures with bodies that in some measure emulated the conditions of their former aqueous homes. The truth of this is evident to this very day, for the bodies of animals (including those of men) are composed very largely of the medium in which their ancestors once existed, namely water. Indeed even the salinity of the blood approximates that of the early oceans.

Once the bodies of these early creatures had become sufficiently adapted they were able to make considerable headway on land. As yet there were no predatory creatures (these had yet to develop from their own kind), and edible material was abundant among the lush vegetation and great primeval forests that covered the land.

These creatures eventually gave rise to the great reptiles of prehistory, many of which attained truly gargantuan proportions. There were, for example, the brontosaurus measuring 67 feet in length and weighing 25 tons, the 88-foot diplodocus, the great brachiosaurus weighing 50 tons, and the smaller but incredibly ferocious tyrannosaurus (47 feet). Eventually these monstrous lizards died out, disappearing entirely and relatively quickly from the primeval scene. The reasons for their demise remain something of a mystery to this day. In many instances the creatures had become so ponderous that they were no longer really compatible with their environment. The brains of many were extremely small, and for this reason a considerable number must have perished simply by blundering into swamps and marshes. Indeed there are a number of instances in which fossilized remains point strongly to this possibility. There is also a belief that the small and developing mammals located in the reptiles' habitat ate or otherwise destroyed the reptiles' eggs. However, the main cause of the demise of these living juggernauts was most probably the advent of radical changes in climate and geology.

For whatever reason, mammals finally became predominant. Eventually there appeared monkey- and ape-like creatures. These lived mainly in the trees, but when climatic changes caused large-scale deforestation they were compelled to live at ground level where conditions were more perilous. Their lot was made worse by the advent of several ice ages, which necessitated large-scale migration. From the higher apes, man gradually evolved in several stages, the remains of which have been found in various parts of the world.

This then, very briefly, is the pattern of evolutionary progress here on Earth, the pattern of one planet out of a numberless host. Our theme, however, is not the pattern of life on Earth but of communication with alien races separated from us by great gulfs of time and distance. Therefore our interest in life processes and forms must be wider.

We know only the life of Earth. This remains a self-evident truth, and until we have some kind of contact with aliens it is a circumstance that cannot change. It is still possible, however, to reflect to some extent on the forms of alien life. We must remember that the laws of matter remain valid throughout the universe and that evolving life anywhere is bound by them. The deeper implications of this fact are frequently ignored by science-fiction writers whose basic precept at times seems to be that anything is possible! So far as life forms are concerned this is most unlikely.

Life as we know it is based on the three states of matter: solid (carbon compounds), liquid (water), and gaseous (oxygen). Carbon is eminently well suited. It has a chemical valence of 4, enabling it to combine freely with other elements which results in a host of varied compounds. There is virtually no other element with this enormous potential, and it is unlikely that life anywhere else will be based on an element other than carbon. Carbon has the added advantage of low molecular weight and the ability to form a *gaseous* oxide with the oxygen of the atmosphere. Silicon, whose valence is also 4, has sometimes been suggested as an alternative to carbon in this respect. Unfortunately silicon is disobliging enough to form a *solid* oxide with oxygen. We are, therefore, left with the unlikely (and presumably painful) process of silicon-based creatures exhaling quartz! In general silicon compounds are much less appropriate in a biochemical context.

Water, in spite of its abundance, is quite a remarkable compound. Its freezing and boiling points are both relatively high (0° C and 100° C), which means that the range of temperature at which it can exist as a liquid (100°) is considerable. Indeed in this respect it is exceeded by only one other compound, hydrogen fluoride. Other noteworthy features about water are its high specific heat, surface tension, and so on. Its solvent powers are of particular significance.

It is therefore difficult to conceive of any other combination of materials more likely than water to lead to the establishment of higher forms of life. The chances of such forms occurring are perhaps not entirely non-existent, and it remains fundamentally important that we do not close our minds completely to other possibilities. However, it seems much more reasonable to expect nature to use the materials that are most suitable.

When we stop to consider all the teeming life on our own planet, it is impossible not to be struck by the enormous diversity of its forms. There are great creatures that live in the oceans, all manner of animals that exist on land, and forms of life that can fly. There are insects in enormous profusion, and there are snakes and a host of other reptiles. Above all there is man, a thinking, rational creature, who has created a society of enormous complexity and potential.

Just why should life have shown such diversity of form? The external characteristics of a living creature are largely dictated by the circumstances and environment in which its remote predecessors lived. In the initial stages of evolution, living creatures as they grew larger were compelled to increase their external surface area in order to obtain the nourishment they required. Although this was satisfactory in the case of primitive organisms, there were obvious disadvantages in the case of higher and more advanced life forms. In the latter case it was more convenient to increase the surface area of internal organs, which explains why the total area of lungs, intestines, kidneys, and so on are considerably greater in extent than that of the external skin area. This also explains why it is unlikely that birds or insects can ever exceed the higher animals in size.

The question of intelligence must also be examined. For intelligence to be of a high order the brain must possess a certain minimum mass. Such a brain must be adequately 'fed'. In order to achieve this a body of certain minimum mass is also essential – a mass in this case considerably greater than that of the corresponding brain. This is why small animals cannot possibly be as intelligent as man. It also makes things very difficult for the disembodied brains that have a habit of appearing in the pages of science fiction!

These facts should not, however, be misinterpreted. The large mass of the elephant does not imply greater intelligence than that of man, for the *form* of the brain is also of importance, that is, as well as the brain's relative size such factors as the number of nerve connections must be considered. The classic case of the large dinosaurs can be used as an example here. The brain of the great diplodocus, for example, was incredibly small and ineffective.

We saw in the preceding chapter that it was necessary for the stable period of a star's life to exceed by a considerable margin the time required for life initiation and development. The latter is an incredibly slow process, and in the case of Earth something of the order of 3,000 million years had to pass before intelligent life appeared on the scene. Earlier in this chapter we considered some of the possible initiating mechanisms. It is virtually certain that nature did not hit upon the right combination in the very first instance, and we must conclude that countless millions of 'experiments' were necessary before the first primitive organism survived and began to develop.

The period of time it takes for intelligent life to appear on other planets might be either shorter or longer than on Earth, depending upon the prevailing conditions. The most critical factor here is probably that of temperature. A warmer planet than Earth could be expected to have a shorter period for life initiation by virtue of the greater reactivity of the chemical compounds involved. For a colder planet the converse would be equally true.

In the preceding chapter we also discussed the question of atmospheric retention in relation to planetary mass. The latter would be of prime importance in deciding the form that living beings might assume. Because of its relatively low gravity a small, light planet could reasonably be expected to produce either creatures of considerable bulk or tall and relatively fragile beings. A planet of this nature could also suffer from atmosphere dissipation resulting in the provision of large chests and extensive lungs for its occupants. Beings on a large, fairly massive planet might be squat and fairly small. Here the prospect of intelligent quadrupeds cannot be overlooked.

Table 5 sums up very briefly just a few of the intriguing possibilities.

Type of Planet	*Probable Life Form*
Small mass, low gravity, thin attenuated atmosphere	Tall, large chests, large nostrils (See Figure 7a.)
Large mass, fairly high gravity, dense atmosphere	Small, squat, quadruped, powerful muscles (See Figure 7b.)
Medium mass, average gravity, moderate atmosphere	Similar in some respects to terrestrial life forms (See Figure 7c.)

TABLE 5

A table of this sort could very easily be extended to cover planets of a more unorthodox and perhaps even bizarre nature, that is, planets that are wholly covered by water or are largely swamp. However, since our essential theme is that of contacting alien civilizations, such instances are hardly relevant. Presumably an entirely water-covered planet *could* lead to an ichthyological type of civilization complete with a submarine society and technology, but the prospects for this seem considerably more remote. (See Figure 8a and Figure 8b.)

The possibility of intelligent forms of life having the natural ability to fly cannot be entirely ignored. To our terrestrial minds this may seem like a fantastic concept but closer examination reveals that this need not necessarily be so. On Earth, the ratio

Figure 7a. Beings on a small, thin atmosphere, low-gravity planet. Because of these conditions they possess enhanced lung capacity, perhaps even gills, and are tall and of rather fragile skeletal structure. Artist: Ronald Russell.

Figure 7b. Form of creature which might inhabit a heavy-gravity planet containing an atmosphere of erosive gases. Artist: Nick Cudworth. (Reproduced from *ICI Magazine* by permission of ICI Ltd.)

Figure 7c. Intelligent bipeds of a world very similar to Earth. Artist: Ronald Russell.

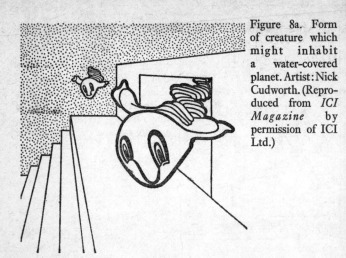

Figure 8a. Form of creature which might inhabit a water-covered planet. Artist: Nick Cudworth. (Reproduced from *ICI Magazine* by permission of ICI Ltd.)

Figure 8b. Intelligent fish-like beings on a water-covered world. Artist: Ronald Russell.

between atmospheric density and gravity is too low to permit natural, unaided flight other than by birds – creatures that are in no instance really large. In the case of planets where this ratio is high the possibility does exist that intelligent beings might just have been so endowed and could therefore be similar to birds in several respects. (See Figure 20d, page 190.) A civilized society enjoying the gift of flying would be a most fascinating one, and to imagine it we have only to visualize the possibilities for our own civilization in such circumstances.

The idea of intelligent plant life has often occurred to writers of science fiction. Again this seems ridiculous until we consider the matter more closely. A planet orbiting a hotter, more luminous sun than ours and having an atmosphere rich in carbon dioxide could provide the necessary environment. We normally associate plants with a fixed existence in one spot, virtually immobile. A plant must obtain its essential salts from the ground via its roots and its carbon dioxide from the air via its leaves. It is not impossible, however, that an alternative arrangement dispensing with permanent fixed roots could have evolved – perhaps a system whereby temporary roots or suckers could be driven into the ground and removed at will. Presumably some means of locomotion for such a plant could also have evolved. As with an intelligent submarine race, however, the prospects for a civilization composed of intelligent plant life seem much more remote. The forms of the creatures with whom we may eventually make contact may differ greatly from ourselves, but generally speaking it seems improbable that they will be fish-like, insect-like, or vegetable – but, we must hasten to add, not impossible!

Though there may be planets resembling our own in nearly all respects it does not follow that evolution on these planets will have followed precisely the same path. On Earth evolution was undoubtedly influenced by such factors as climatic change (e.g., ice ages, spells of heat), geological shifts, and so on. Such events will also have occurred on other terrestrial-type planets but on a different time scale. The effect of these events on races at differing periods of evolution must inevitably lead to differences in the biological 'end-product'. Indeed, differing conditions at the time of life initiation could result in the evolution of entirely alternative forms of life.

Even completely terrestrial conditions can be no guarantee that life on the planet concerned will be totally like our own, for chance is always likely to play a part. Although the body of *Homo sapiens* is entirely suitable in most respects, there are anomalies that may be attributed more to this element of chance than to strict evolutionary development. It may be beneficial if we take a closer look at some of these anomalies.

It is obviously fitting to have the senses controlled and co-ordinated by one compact brain. The brain lies in close proximity to the eyes, ears, mouth, and nose, that is, to the organs of sight, hearing, speech, taste, and smell. In the case of creatures travelling on all fours the brain is also conveniently close to the feet and the legs. When, however, the early ape determined that it could travel about on two legs equally well, thus freeing the other two for other duties, the brain was at once rendered more remote from the means of locomotion, a position in some respects less favourable. No one will fault the location of the eyes and the ears in this elevated position, although it has been suggested that the olfactory organs would be more satisfactory at a lower level.

In terrestrial creatures the mouth serves a dual purpose. It is used for eating and drinking as well as for conveying audible intelligence. Certain other organs have also been made to serve a dual function, for which there is no apparent reason. The possibility of alien creatures possessing separate organs for these functions remains.

On Earth the number of limbs given to creatures varies. The snake has none, whereas some insects have several hundred. Animals and birds have four – in the case of birds two limbs are wings. Animals generally move on all fours, but the ape family and man use two of them for alternative purposes. Earlier we spoke of the possibility of an intelligent quadruped civilization existing on a high mass/heavy gravity planet. It might well be asked at this point whether another pair of limbs that would function as arms might not have developed, since creatures requiring all four limbs for purposes of locomotion would undoubtedly be inhibited in their passage towards an advanced technological civilization. (See Figure 20c, p. 189.)

Clearly there are a considerable number of feasible variations even on terrestrial-type planets. This trend must be infinitely more pronounced on planets where conditions that are favourable to life are nevertheless markedly dissimilar from those on Earth. We are thus faced with the very real possibility that there exist highly intelligent races in advanced civilizations whose external characteristics are to us quite simply non-human. Such creatures if they exist are certainly aliens but they are *not* monsters. They are beings *in their own right*. If to us *they* are monsters (merely because they are totally different) then to them *we* are monsters. This is an inescapable fact no matter how much we may dislike it or seek to argue round it.

The theme of extraterrestrial life is obviously one about which much can be said. The possibilities are intriguing, fascinating, and mysterious. We cannot but be reminded of the famous words of Alexander Pope, in his 'Essay on Man':

> Observe how system into system runs,
> What other planets circle other suns,
> What varied Beings people every star.

Are the first living cells even now being created on a world of Beta Centauri? Have the first great reptiles appeared under the sun we call Epsilon Indi? Does primitive man even now peer out, dazzled by the glare of Tau Ceti? Is there a creature of massive intellect beneath the twin suns of Antares sending a stream of electronic intelligence towards our sun? We do not know, but the days are gone for ever when science would categorically declare these things to be impossible.

6 Antenna Versus Starship

IN THE preface we sought to allot to interstellar communication its own particular region in the 'spectrum' of our cosmic future. We can say that we have just entered the era of interplanetary travel and if asked for the specific date of that entry we would almost certainly have to say 4 October 1957, the day on which the first artificial space satellite, Sputnik I, was sent into orbit round the earth. It is obvious that as yet we have advanced no great distance into the first of our astronautical epochs. (See Figure 9.) Nevertheless what has been achieved so

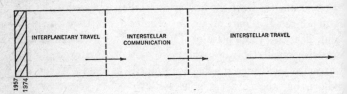

Figure 9

far is quite remarkable. We must stress of course that the diagram in Figure 9 is only a crude yardstick and should certainly not be taken too literally. Clearly the era of purely interplanetary travel will not abruptly terminate with the advent of interstellar communication. Neither does the dotted line denoting the commencement of interstellar travel imply the immediate cessation of all forms of interstellar electronic communication. The new epochs are merely superimposed on the older ones, which will in each case continue to develop. Although we have set the beginnings of interstellar communication some way ahead, it could justifiably be argued that the work already carried out by Dr Frank Drake and his colleagues in Project Ozma marks the commencement of this particular era.

However, since this was merely a brief, desultory experiment it is probably advisable to regard the real start of this period as the day on which permanent programmes of experimental work are launched.

It should also be fairly obvious that no estimated date can be given to the beginnings of these two future epochs. It would be a most wildly sanguine person indeed who would care to hazard a guess when the genesis of interstellar travel will be, though there are those who maintain with some justification that this, too, began on 4 October 1957. These people see interstellar travel merely as an extension of the basic astronautical theme. Nevertheless interstellar travel introduces wholly new parameters of time, distance, and power and is to all intents and purposes an entirely different conception. Indeed it might be said to represent a totally new dimension.

Returning for a moment to our diagram we see that beyond the era of interstellar travel no further epoch is shown. We simply dare not look that far ahead. It would be easy and no doubt fascinating to speak of intergalactic travel, of journeys between the various *universes*, but this involves further parameters of time and distance that are even more awesome and frightening.

All we have sought to do so far is to give interstellar communication a proper frame of reference, to see it not as a separate entity but as a further logical step in the process of coming to terms with the known universe. If it should be asked why interstellar communication should *precede* the coming of interstellar travel it is necessary that the pros and cons relating to each be examined. This, then, is the aim of our present chapter and the reason for its apparently paradoxical title.

Interstellar travel is more than just another new concept. It is the greatest and most exciting concept ever to confront mankind. It represents a dream that has tantalized past generations and will exert the same incredible fascination for countless generations yet unborn. Though electronic communication between ourselves and the peoples of the stars must produce great problems these will be nothing compared to those involved in any attempt to cross the time and distance barrier physically.

Over the years many sincere men have sought to convince us that travel beyond the confines of the solar system is an unrealizable dream. There are, however, also those who refuse to surrender to what they think is a negative attitude. From the outset it must be made very plain that interstellar travel *will* bring problems of monumental extent, of frightening magnitude. This point we *must* concede.

Just what exactly are the essential facets of the problem? The nearest star to the sun is Alpha Centauri lying 4·3 light years distant. Compared to the more remote stars and the inconceivably distant island universes such a distance may seem very short indeed. In a strictly relative sense this is true. At the same time we must remember that the small matter of 4·3 light years separating us from Alpha Centauri is equivalent to no less than 25 million, million miles. A spaceship travelling at 120,000 miles per hour, a considerable velocity, would take 24,000 years to reach this particular star!

If it were possible to devise a ship capable of travelling very close to the speed of light (186,000 miles per second), the position would look considerably more promising. In about ten terrestrial years the round trip could be completed (allowing for the necessary acceleration and deceleration periods). But even here the cold logistics relating to food, water, air, and fuel are alarming. The so-called time-dilation effect that follows from some of the equations of Einstein's Relativity Theory would certainly work appreciably in our favour. Despite this, many other serious problems remain.

A space-rocket as it leaves Earth has an initial mass that is made up of propellant and payload. The propellant is responsible for imparting to the rocket its final velocity. Once exhausted only the payload, or final mass, remains.

Now let us delve a little more deeply into the mechanics of the situation. A rocket's performance depends for the most part on the velocity of the gases emitted from its tubes (i.e., its exhaust). Let us call this V_t. The final or maximum velocity that it attains will be V_m. Initial mass, that is, payload and propellant, can be denoted by M_I and payload alone by M_F.

Now let us have a look at the simple mathematical expression that links these factors. This is

$$\frac{M_I}{M_F} = \left[\frac{C + V_m}{C - V_m}\right]^{\frac{C}{2V_t}}$$

(1) If V_m tends towards the speed of light (C) then numerator $(C + V_m)$ becomes very large and denominator $C - V_m$ very small.

(2) If V_t is low then $\frac{C}{2V_t}$ becomes very large.

The combined result of (1) and (2) is that the entire right-hand side of the equation becomes very large.

$$\text{i.e., } \frac{M_I}{M_F} \text{ is large,}$$

$\therefore M_F$ is very small or M_I very large.

Thus *low* exhaust velocity means large initial mass, and low final mass, that is, lots of propellant and not much payload. The essential requirement is therefore a propellant possessing very high exhaust velocity – in other words, *less* propellant and much *higher* payload.

If hydrogen is wholly converted to helium, our tube or exhaust velocity (V_t) is approximately equal to $\frac{C}{8}$, that is, one-eighth of the velocity of light, or 23,250 miles per second. Our final (maximum) velocity we will set at 0·99C (i.e., almost the speed of light) when the ratio $\frac{M_I}{M_F}$ or initial mass to final mass becomes $1,600 \times 10^6$ (1,600 million). This is still not good enough for it means our initial mass would amount to 1,600 million times the final mass.

Unless we resort to the use of matter and antimatter – rather like a science-fiction concept – we cannot hope to improve on this apparently hopeless state of affairs.

So far, however, we have been thinking in terms of a journey that could be accomplished well within the normal human life span. The idea of 'generation migration' has long been preferred by serious protagonists of interstellar flight. (See illustrated section). This relatively more realistic concept envisages the dispatch from the solar system of a vast starship containing a number of genetically sound young men and young women. In the normal course of events a new generation is born, a process repeated many times during the long years in space. Eventually the distant descendants of the first stellar astronauts reach the star system of their forebears' choice. Writing in the *Journal of the British Interplanetary Society* in 1952, Dr L. R. Shepherd said:

> It would be as though the vessel had set out for its destination under the command of King Canute and arrived with President Truman in control. The original crew would be legendary figures in the minds of those who finally came to the new world. Between them would lie the drama of perhaps 10,000 souls who had been born and had lived and died in an alien 'world' without knowing a natural home.
>
> Thus interstellar exploration and colonization may be vastly different from the exploration and colonization of our own world or even of our own system. It may require a revolution in our way of life not only socially but biologically if we are ever to become a galactic people.

Obviously the problems here, moral and sociological as well as technical, are immense. Nevertheless it seems not at all unlikely that, once a journey of this nature becomes a feasible proposition, man will be unable to resist its massive challenge. If the proposition is made, it will be one quite without parallel in the long and exciting history of our species.

The peculiar circumstances of generation star travel raise ethical as well as technical problems. Within the thin metal walls of the artificial planetoid, human beings would be conceived whose destiny had been wholly preordained. All the mid-generations would be mere links in a chain of purpose, the means to an end. The strictest control of marriage and birth rate would be utterly essential – and this is only one aspect. Indeed the entire migrating community and all its descendants would be subject to a measure of discipline and control without equal in history.

Despite all these very massive problems that never dare be underestimated, the concept of generation travel allows us to circumvent the technical impasse we came up against earlier. Therefore it should prove interesting to re-examine and re-appraise the issue in the light of this.

We are now able to think in terms of much lower velocities. Let us for example consider a journey to Alpha Centauri at one-fiftieth of the speed of light (i.e., $\frac{C}{50}$). Going back to our original mathematical expression the following interesting tale unfolds:

$$\frac{M_I}{M_F} = \left[\frac{C + V_m}{C - V_m}\right]^{\frac{C}{2V_t}} \quad \text{N.B. } V_t = \frac{C}{8} \text{ (perfect nuclear fusion)}$$

$$\text{Thus:} \frac{M_I}{M_F} = \left[\frac{C + \dfrac{C}{50}}{C - \dfrac{C}{50}}\right]^{\frac{c}{2Vt}} = \left[\frac{51}{49}\right]^4 = 1 \cdot 17$$

$$\text{i.e.,} \quad \frac{M_I}{M_F} = 1 \cdot 17$$

$$\therefore M_I = 1 \cdot 17 \, M_F$$

If, therefore, the final mass of the starship were to be 5,000 tons, initial mass would be required to be just a little short of 6,000 tons. Although this is a journey that would occupy a period in excess of two centuries, the mathematics are much more in our favour.

Earlier we referred briefly to what is now popularly known as relativistic time dilation. Were it possible to convert matter entirely into energy, velocities *approaching* that of light might be realized. Under these conditions time dilation should become apparent and therefore materially reduce the period required for voyages to the stars. A clock aboard a starship would then run 'slow' compared to one on Earth, while a corresponding retardation in the biological, physical, and chemical processes in the bodies of the ship's occupants could be anticipated. In other words the ageing process would slow down. Although this may seem to be an incredibly fantastic idea, research into certain subatomic particles that move at almost light velocity has indicated that it has a certain validity.

There is a tendency to extrapolate the ever-rising speeds achieved by mankind and thereby 'prove' that in a certain time the necessary velocities to achieve interstellar flight will be attained. It may be of interest to look a little more closely at this exercise, the mechanics of which are as easy as they are intriguing. One merely plots the incremental rise in the speed of ships, trains, cars, and aircraft over a number of years. As a result each form of transport provides a distinctive short curve. A line is then drawn touching each curve tangentially at its highest point. This line, known as an exponential curve, continues upwards enabling (or apparently enabling) us to predict future attainable velocities. By A.D. 2100, for example, it can be shown that 1 per cent of the speed of light should be possible, rising to 5 per cent a century later.

So far this form of prediction has proved remarkably accurate, and it is thought that over the next two or three decades it will remain a fairly valid pointer. Whether it will hold a century or two from now is more debatable. Such a curve is only a mathematical expression, not an irrefutable oracle. Other factors invariably intrude, and as a result the graph is modified out of all recognition. Indeed an exponential cannot have indefinite validity and must eventually either level off or reverse. Such a curve is termed a *sigmoid*. In their initial stages exponentials and sigmoids are virtually one and the same. Both double in value in roughly the same time, and divergence, when it begins, is gradual.

The weakness of an exponential can be easily demonstrated in our present context if we consider the velocity of light. In this case the speed exponential not only reaches this value but climbs *beyond* it – an absurdity, in view of Einstein. Theoretically, however, a sigmoid would reach a value approximately half the velocity of light in the same time. Thereafter it would be expected to level off, though perhaps attaining a value *very close* to that of light velocity. According to Einstein such a value represents the approximate limit we could hope to reach.

Those who favour the exponential outlook frequently point to new modes of propulsion modifying the curve in their favour. They demonstrate the sharply rising angle given to the curve first by aircraft and then by jet and rocket, and they point to probable new propulsion techniques (e.g., ion and plasma systems) continuing the process.

There is unfortunately a flaw in this argument. The possibilities inherent in reaction thrust (i.e., jet and rocket) were well known long before systems embodying it were perfected. Only delays in metallurgical 'know-how' retarded the advent of these power sources. Though new and revolutionary propulsive techniques are now being considered it would be incorrect to say that the full possibilities and potential of these techniques are clearly understood. In other words they might and probably will steepen the exponential curve, but to an as yet undetermined extent.

However, as events stand, the velocity exponential curve remains encouraging and, so long as it does not assume sigmoidial form unexpectedly, 1 per cent of light velocity should be attained around A.D. 2100 and 5 per cent about A.D. 2200. These tremendous velocities (1,863 miles/second and 9,315 miles/second) certainly bring interstellar travel to the earliest thresholds of possibility.

Interstellar space travel is not therefore impossible. At present, however, all we can do is try to assess the basic requirements for its accomplishment. Travel to the stars has long been a subject possessing vociferous adherents – and equally vociferous opponents. One hundred years ago the idea of lunar and interplanetary journeys was regarded as the sheerest fantasy (we might even say heresy). At that point in human affairs the state of our skills and degree of our knowledge rendered this inevitable. Two or three centuries from now our descendants will in all probability look back and draw a similar parallel. Amazing new sources of energy and revolutionary techniques may by then have placed the key to the stars in the hands of our race. Surely man who has come so far, who has progressed from the Stone Age sling to the moon rocket, will not be stopped for ever on the shores of the dark and mysterious interstellar ocean that lies eternally waiting beyond the farthest planet of the solar system.

Regrettably a great deal of the more lurid type of science fiction has done much to discredit the concept of travel to the stars. The sort of space opera that has hero and heroine roaming the galaxy at will, dressed in the most unlikely (and frequently revealing) of spacesuits is no doubt highly entertaining. Unfortunately it bears little or no relation to the realities of the subject. Tales of the distant future that tell of the difficulties, the doubts, the great perils, and incredible poignancies of interstellar journeyings are to be welcomed. These tales place the thing where it rightly belongs – at a remote point in the history of our race and of our planet. We see it as an ultimate destiny, something worth striving after. But the light-hearted adventures, amorous and otherwise, of husky spacemen and attractive young spacewomen could surely just as easily take place on the moon, Mars, or the more appropriate of the asteroids (which is probably what will happen anyway!) Must they cavort about on the worlds of Alpha Centauri, Tau Ceti, and Epsilon Indi? It seems such a long way to go!

Whether we believe in the feasibility of transgalactic journeys or not, the fact persists that interstellar travel at the moment and throughout the foreseeable future must remain an interesting, fascinating, but entirely academic question. What therefore can we do now or in the near future towards establishing a link with some of our fellow beings within the Milky Way? The words 'fellow beings' are used quite deliberately. Their form may be vastly different from our own, perhaps in some instances even grotesque or repulsive. Nevertheless if these are intelligent, reasoning creatures they deserve recognition as such, irrespective of appearance. A different form should not, indeed must not, prevent our extending the hand of friendship towards them – or refusing to grasp (metaphorically at present!) the hand, paw, or claw of friendship held out to us.

What we *can* do *now* is endeavour to transmit intelligible signals to these peoples and, if possible, to receive theirs. It is a small, first step but a real and significant one.

It has been estimated that a ten-word electronic message could be transmitted over a gulf of 12 light years (73 million, million miles) for an expenditure of energy to the value of a few pounds. Obviously the equipment to achieve this would be rather more expensive! Yet even this expenditure would be modest indeed compared with that necessary to send men and women into the dark recesses of interstellar space.

To transmit a signal over such a vast distance is not a particularly involved business. The pulses of energy radiate outwards, suffering only the loss in strength due to what is known as the inverse square law. This is merely an elaborate way of saying that the further such energy has to travel the more attenuated it will become.

If S represents the signal strength and d the distance involved then

$$S \propto \frac{1}{d^2} \;(\propto = \text{proportional to})$$
or
$$S = \frac{k}{d^2} \;(\text{where k is a constant})$$

Since d is raised to the power 2, signal strength will obviously decrease significantly with increasing distance.

These pulses, though relatively weak when they leave the

transmitting antenna, can nevertheless be detected and adequately amplified in suitable receiving apparatus. Readers familiar with the essentials of amateur short-wave radio transmission will readily appreciate this point. Using very modest transmitting equipment it is not at all difficult to radiate a signal capable of being picked up with ease on the other side of the globe by receiving apparatus even more modest.

Thus we can see how *relatively* easy it is to travel many light years on the invisible wings of radio – and how incredibly difficult it is going to be to reach out via a starship even as 'short' a distance as $4\frac{1}{4}$ light years. We must repeat, however, that this does not mean interstellar communication will not bring a full quota of problems – or that interstellar travel will for ever be impossible. The former will call for infinite patience, the latter for infinite courage!

An issue that has been raised on a number of occasions concerns the topics we might discuss with alien star peoples once a satisfactory medium of communication and understanding has been attained. There is so very much we would want to learn about these peoples, their worlds, their histories, their outlooks, that it might be extremely difficult to know where to start. Consider the priceless knowledge that an advanced and cultured race 500 or 1,000 years our elders might bestow on us. A contact of this nature could well represent the coming of a great millennium, a new renaissance that might mean the salvation of our planet and its peoples. It could provide us with the answer to problems that have sorely beset mankind for generations – how to live together amicably; how to feed Earth's starving millions; how to treat the dread diseases that have so far baffled medicine. New sources of energy might be revealed to us, a new technology created. We might even be given the key enabling us to reach the stars. There is, of course, an inherent risk in all this. Would certain of such gifts be used in a manner for which they were not intended? Would the secret of a new and perhaps limitless source of energy be put to the wrong use? Would a gift from the star peoples be translated into the bomb to end all bombs? We cannot tell. Perhaps, however, a new philosophy of humanity will also reach us from the stars.

Will electronic forms of communications be enough? Will men be content merely to speak to alien peoples and to *learn* of their planets and affairs? Will there not arise an infinite longing to see these worlds, to survey their great works? This seems not at all unlikely. Perhaps an inexpressible desire for closer, more real contacts will materially speed the day when man will rend asunder the bonds of time and distance. The dreams of today may be the irresistible driving force of tomorrow.

7 The Stars Have Voices

BEFORE examining the techniques whereby intelligent signals may one day be received from space, time should perhaps be spent looking at the history and fundamentals of radioastronomy. In this chapter we will concern ourselves not with artificial signals emanating from civilizations on alien worlds but with the natural, spontaneous babble and chatter of the stars themselves.

In 1873 James Clerk-Maxwell showed that light waves were not the only form of electromagnetic radiation, and he predicted mathematically the existence of wavelengths ranged on either side of the optical portion of the spectrum. Fourteen years later Heinrich Hertz discovered radio waves, for a time known as Hertzian waves in recognition of him.

Although radioastronomy is regarded as a relative newcomer to the scene, the *enfant terrible* of modern astronomy, its origins go back almost to the beginnings of radio itself. In 1890 Arthur Edwin Kennelly, prompted by Thomas Edison, suggested in a letter to the principal of the Lick Observatory in California that since light and heat radiation are both received from the sun, radiations of longer wavelength might also be emanating from that body and that these might not only be detected but picked up in such a way as to render them audible. Edison carried out some experimental work in this respect that though ingeniously conceived could not, we now realize, have been successful. In 1894 Sir Oliver Lodge also carried out a number of unsuccessful experiments.

We must jump nearly four decades to find the real genesis of radioastronomy. In 1932 Karl Jansky, a young engineer, was carrying out some experiments on behalf of the Bell Laboratories. The purpose of these experiments was to determine the noise level likely to be encountered in a sensitive short-wave receiver designed for long-distance reception if it was used in conjunction with a directional aerial. Jansky's aerial system, even though it was only a crude affair mounted on a pair of old car wheels, was able to scan the entire sky. The noises he most expected were those caused by electrical storms and other static disturbances, which generally show up as harsh intermittent crackles. What he heard instead was a steady hissing sound. To his surprise he noticed that this sound was at its loudest when the antenna was directed towards a particular region of the sky, a region coinciding with that portion of the Milky Way located in the constellation Sagittarius (see illustrated section). It soon became obvious that what Jansky was hearing was nothing less than radiation originating *in* the galaxy!

At the time the news was well publicized, especially in the United States where one radio station even went so far as to broadcast 'the sound of the stars' coming from Jansky's receiver. Surprisingly, this important discovery raised only a mild ripple of interest in scientific circles where the strange radiation was regarded more as a curiosity than the valuable adjunct to cosmological research it was destined to become. There was one man, however, who did appreciate the possibilities. This was an amateur radio operator named Grote Reber to whom the distinction of being the world's first radioastronomer must go. Indeed, Reber for a number of years was the world's only one! In the garden of his home in Illinois he constructed a steerable parabolic reflector 30 feet in diameter – the first real radio-telescope. Reber used this in conjunction with sensitive receiving apparatus.

Although his first trials were not particularly successful, he was eventually able to detect at a wavelength of 60 centimetres similar radiation to that which Jansky had discovered at 15 metres. His equipment being considerably more sophisticated than Jansky's, he was able to draw up charts showing in fair detail the points from which these radio waves came. Surprisingly, the origin of the radiation proved to lie in regions where no visible bodies were in evidence.

Reber's work was published during the early part of the Second World War, and since the world was then preoccupied with more immediate and compelling problems his work did not receive the recognition it deserved. Fortunately the work he had done was not forgotten. When the war finally ended,

Reber's results began to attract considerable attention. Interest in the subject had been further fostered as a result of certain radar research carried out during the war which showed that the sun transmitted radiation having a wavelength of one metre. It had also been discovered that meteor trails were responsible for radar echoes.

Reber's work in drawing up a radio chart of the sky was augmented after the war by work done by J. C. Hey, who had been actively engaged in radar research during the war. Using a vastly improved antenna Hey drew up a comprehensive map of the northern sky. This continued to portray the Milky Way as the most prominent radio feature, although a point in the constellation Cygnus also proved to be a powerful source of radio waves.

Both branches of astronomy (radioastronomy and optical astronomy) have a common basis in that electromagnetic radiation from the heavens is involved in each case. There is, however, a distinct difference, since objects in the sky responsible for the emission of radio waves generally bear little semblance to objects of the visible universe. We cannot, for example, see a transparent mass of hydrogen gas because this does not emit radiation in the visible portion of the spectrum. It may, however, emit radiation of a radio frequency that can be detected by a radio-telescope. We can therefore in a sense 'see' the hydrogen cloud, for we are now aware of its presence, and its contours and limits can be charted.

We might conveniently liken the position to a system of 'windows' in the sky. The diagram in Figure 10 illustrates fairly clearly what is meant.

The electromagnetic spectrum, although extensive, is comprised mostly of radiation that is unable to penetrate the ionosphere and atmosphere of our planet. The ionosphere, which is essentially an invisible concentric shell of electrons lying a hundred miles or so above the Earth's surface, is extremely useful and without it long-distance radio communication would be virtually impossible. It reflects the longer wavelengths (i.e., broadcast signals) back to Earth making it possible for listeners in Europe to hear programmes broadcast from points as far distant as the high Andes or the surf-bound islands of the Pacific.

Occasionally the ionosphere becomes non-cooperative, and when this happens signals from transmitters become 'lost'.

Instead of being reflected back they are absorbed entirely by this layer. The ionosphere reacts to incoming cosmic radiation just as it does to artificial radiation coming from radio transmitters. Much is absorbed and cannot therefore pass through to Earth, whereas some, by virtue of total reflection, is turned back. Reference to the diagram in Figure 10 shows how at the lower

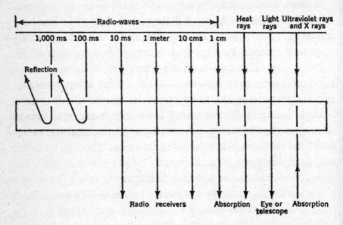

Figure 10

end of the spectrum incoming waves are reflected by the ionosphere and how, towards the upper end of the spectrum, gamma rays, X rays, ultraviolet rays, and infrared rays are completely absorbed. In only two regions of the spectrum can radiation reach Earth's surface – the two so-called optical and radio 'windows'. In neither window is the appropriate radiation absorbed or reflected by an atmospheric layer. We can thus either see the various celestial bodies or become aware of their presence by means of radio-telescopes.

Nights in which the sky is shrouded by clouds or mist are disappointing to the optical astronomer. On these occasions the most powerful and expensive telescopes in the world with all their sophisticated auxiliary equipment remain useless behind tightly closed domes. In other words one of the two windows into space has been temporarily closed. The radio window, however, is open just as wide, and even though the great 'dish' of a radio-telescope may be enveloped in thick fog or streaming water from thick rain clouds it can 'see' the heavens with perfect clarity. The heavens it scans, however, are apparently very different from those that its optical counterpart would see were the skies free from clouds.

Let us assume for a moment that some miraculous process enables our eyes to respond to radiation in the radio portion of the spectrum. At once the sky becomes utterly alien. The sun is now considerably less brilliant, while the Milky Way appears as a very bright band of light girdling the sky. Gone are all the old familiar constellations and well-known stars. In their place have come new configurations and bright bodies that are complete strangers to us. The moon and planets are barely visible, although the great planet Jupiter emits such vivid flashes of light that it is as if an enormous electrical storm were raging over its entire surface. Nevertheless this is the selfsame sky that we formerly scanned with our normal 'optical' eyes. Nothing has really changed. It has always been this way. As we would expect, objects emitting strong visible radiation appear faint. Conversely, objects emitting intense radio radiation but little or nothing else are now seen as very bright objects.

These objects have been termed *radio stars*, and it is obvious that they must be very different from ordinary stars. The first of these to be discovered lies in the constellation of Cygnus, the Swan. However, it is not a member of our galaxy, for it is located well outside the Milky Way. (See illustrated section.)

The most powerful of all the so-called radio stars is to be found in the well-known northern constellation of Cassiopeia. Another particularly strong radio star lies in the constellation of Taurus, the Bull, and it is thought to represent the remains of a supernova that shone in our skies back in the year A.D. 1054. This radio star is known to optical astronomers as the Crab Nebula on account of its vaguely crab-like appearance. For two years this star shone so brilliantly that it was plainly visible by day! Through an ordinary telescope the Crab Nebula appears to be a very small blue-coloured star surrounded by a cloud of gas. Some idea of the tremendous power of this cosmic convulsion (the supernova of A.D. 1054) may be gained from the fact that the shell of expelled matter is still hurtling outwards at the prodigious speed of 800 miles per second. It is from this shell that the radio signals chiefly emanate.

By now several thousand of these radio stars have been discovered but only a few apparently belong to our galaxy. At first the tendency was to attribute their origin to supernovae. Certainly the remains of the supernovae that had appeared in Taurus, Opiuchus, and Cassiopeia are all sources of radio radiation. Now, however, the question is more open, for it appears that a star in the process of creation is almost as powerful a radiator of radio signals as a star in violent death throes. Clear indication of this is afforded by the 'signals' coming from the Great Nebula in Orion (M42), long regarded as a stellar 'crucible' in which stars are being created.

However, it is not only from embryo and dying stars that radiation of this type emanates, for a group of stars known as flare stars also emit on frequencies within the radio portion of the spectrum.

In Chapter 3 we saw how greater insight into the structure and form of the Milky Way had been derived from radioastronomical techniques. Astronomers had long considered our own galaxy to be similar in configuration to the other remote island universes that they were able to see in optical telescopes. Confirmation of this was difficult, however, due to obscuring masses of dust and gas. Although dust constitutes an optical 'blind', this blind becomes an almost open window so far as radio frequencies are concerned. It was possible, therefore, to pick up the radio emanations from the clouds of hydrogen gas within the spiral arms of the Milky Way and thereby chart their position and form. At last radioastronomers were able to confirm what their optical brethren had long believed – that the Milky Way is a galaxy whose structure is akin to that of other great island universes, such as M31 in Andromeda.

A very obvious question, so far as radioastronomy is concerned, is why elemental hydrogen gas, of which there is such an abundance, should emit radiation. All substances as we know are composed of atoms. In the case of gases, atoms move with considerable velocity. If the gas is dense (e.g., air) many millions of atomic collisions occur each second. In the terribly attenuated hydrogen of interstellar space, however, the atoms are so remote from one another that probably each atom collides with another only once in 200 years. But when this collision does occur a pulse of radio energy results. On this basis it might be thought that little energy could be anticipated. It must be remembered, however, that even though the frequency of collision is low there are countless billions of hydrogen atoms in the vast wastes of space. Thus, somewhere, millions of atoms are colliding every moment, and so a steady radiation results. This radiation has a wavelength of 21 centimetres and is detectable in radio receivers as a steady background noise. This wavelength is of major importance when we come to consider the detection of *artificial* radio signals.

Our nearest neighbour galaxy, M31 in Andromeda, is very similar in all respects to the Milky Way. When, however, its 'radio' outlines are drawn, M31 is seen to be very much larger than shown by a conventional telescope. There is a very good reason for this. The entire galaxy is surrounded not only by a shell of globular star clusters but also by a halo of gas. This halo cannot be seen in an optical telescope, but in a radio-telescope its presence is very obvious. In this respect M31 is not unique for the Milky Way is surrounded by a similar halo.

Certain galaxies have halos considerably greater in extent. A classic example lies in the constellation Centaur, 13 million light years distant, which emits radio waves over an area fifty times as great as that of the visible part. This is Centaurus A, the famous radio source.

A consistently powerful source of radio energy cannot always be identified with a visible object. A case in point, as we have already seen, is that of the first radio star to be discovered. This radio star is situated in the constellation Cygnus and is a most powerful radio source. When it was discovered in 1946 there was apparently no visible object in view. It was only in 1951, after some very precise measurements, that the object was finally located by the great Palomar telescope. What the photographic plate revealed was in fact no normal, faint star but a peculiar-looking object vaguely reminiscent of a dumbbell. (See illustrated section.)

The explanation first put forward was rather a dramatic one – that of two galaxies in collision. What of course was envisaged was not millions of stars colliding with one another – stars are so widely separated that two galaxies meeting would simply pass through one another. Masses of gas would collide, however, and the immense resulting buffeting would almost certainly ensure considerable emission of radio energy.

This idea might have remained valid had not more and more of these unique radio sources been discovered. Surely galaxies could not be for ever colliding since these are even more widely separated than individual stars within a galaxy. Clearly, an alternative explanation had to be sought. Martin Ryle of Cambridge postulated that the objects were simply galaxies so remote that even though their radio energy could reach us their light had virtually come to the limit of its range.

More recently the idea of an exploding galaxy has gained prominence in this respect. What, it might well be asked, could possibly cause an entire galaxy to explode? Occasionally astronomers had noticed galaxies from which jet-like protuberances extended. An excellent example of this is M87 in Virgo from which a long stream of matter protrudes. Is it therefore possible for an entire galaxy – a universe of upwards of 100,000 thousand million stars – to explode?

It is believed that originally a great mass of gas condenses at various points into individual stars, thus forming a galaxy. Under certain conditions (most probably towards the centre of the galaxy) gas condenses not into a great number of individual stars but into one superstar of mammoth dimensions that eventually shrinks and collapses upon itself. This, it is thought, is the reason for the jet of matter emanating from M87 and other galaxies. It must be stressed, however, that this represents not an *ex*plosion but an *im*plosion. In many respects, of course, the results are similar. As the star collapses inwards towards its centre some of the matter is diverted from its path and ejected at a high velocity – virtually an explosive effect. It is thought that the collapse of the central 'superstar' takes place when its diameter is almost equal to that of the solar system. After the contraction process, a small, very massive star remains that is brighter than all the rest in the galaxy while the typical high velocity jet of gas reaches out into intergalactic space.

From such an imploding galaxy there must issue a torrent of radio waves, and although the visible effects are relatively short-lived the radio emanations will go on for a considerably longer period. Not all imploding galaxies will be visible, however. Most are so distant that only the radio-telescope will reveal their presence. Thus it has become possible to identify and locate radio objects in the far reaches of the universe – in regions so remote that no optical telescope can ever hope to plumb their depths.

The farther out radio-telescopes reach, the greater apparently are the number of radio objects. Emphasis is therefore added to the idea of an expanding universe resulting from a great initial explosion. When we identify radio stars 1,000 million light years remote we are in effect seeing the universe as it was when it was very much younger. By reaching out sufficiently far, assuming this to be possible, we may eventually reach the very dawn of creation!

Radioastronomy has given to the science of astronomy startling new vistas. Clearly there must be a limit to the size of optical telescopes, and it must be said that these great instruments are fantastically expensive. Technical problems assume massive proportions as optical telescopes grow larger. At present the great 200-inch Hale Reflector of the Mt Palomar Observatory is the biggest telescope in the world. It took over fifteen years to construct this magnificent instrument, which cost six million dollars. A slightly larger telescope (238 inches) is presently under construction in the Soviet Union, and it seems unlikely that there will ever be telescopes greater in aperture than these giants. Radio-telescopes, on the other hand, are cheaper and easier both to construct and maintain. Giant optical telescopes are built primarily to investigate the distant galaxies; radio-telescopes perform a similar function and are also able to reach out farther into the great deeps of space. Although the two types will remain complementary, it is becoming increasingly obvious that if the universe is ever to yield up its most fundamental secrets it will probably be to the radio-telescope and not to the optical variety.

Radio-telescopes are merely radio receivers of a highly specialized nature. In a later chapter the type of radio-telescope used in Project Ozma will be described in some detail. The word radio-telescope generally conjures up visions of a great dish-like object tilted towards the sky. This of course is merely the aerial and is no more a radio-telescope than the familiar television aerial is a receiver. The dish-shaped antenna of a radio-telescope is analogous to the main lens or mirror of an optical instrument.

Radiation from cosmic objects can in fact be received without recourse to vast, expensive aerial systems. Karl Jansky was using a small, crude arrangement when he first picked up the hiss of the Milky Way, and indeed an ordinary TV aerial can be made to serve the same purpose. Such diminutively simple systems are unfortunately totally inadequate in the realm of serious research.

The bigger the aerial the more radio energy it will collect. This, however, is not the only reason for using a large antenna system. If two radio sources lie very close to each other, a radio-telescope with a small aerial will merely 'show' these as *one* single object. An instrument with a medium-sized aerial will reveal the fact that *two* bodies are present but will not be able to separate them. One having a very large aerial system, however, will be capable of revealing the two radio sources as separate entities. The degree to which this can be achieved is termed the *resolving power* of the instrument and is largely dependent on the size and design of the aerial system.

The huge dish-like contrivance is in fact the reflector, the aerial proper being the vertical mast or rod protruding from the centre of the reflector. The principle upon which the aerial operates is illustrated in Figure 11. Radio waves coming from an object in space impinge on the reflector and are brought to a focus on the central aerial rod. Unwanted waves from other portions of the sky are also reflected but these do not focus on the aerial and so these are not picked up. Unfortunately the aerial system of a radio-telescope is not as conveniently selective as this might imply. Waves that happen to strike the rim of the reflector are reflected in all directions. In the circumstances it is inevitable that some will be reflected to the focus. Therefore, stray signals are always picked up to some extent. If however a very large 'dish' is used, unwanted signals coming from the rim are considerably weaker than the desired signals impinging on the dish proper. The radio receiver into which the signals are fed can itself be rendered highly selective as we shall see when we consider the Project Ozma receiver in a later chapter.

The longer the received wavelength the more troublesome do these interfering stray signals become and consequently the more difficult it is to resolve two adjoining radio sources. The remedy is to use a larger dish, which explains why radio-telescopes are generally such gigantic objects. The wavelengths of radio waves are very much longer than those of light, and so optical telescopes

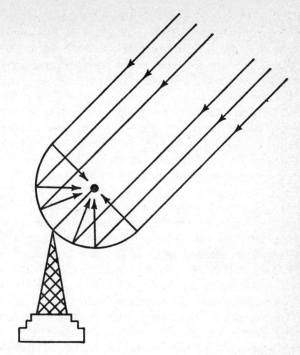

Figure 11

can be smaller and more compact than their radio counterparts. In the resolution of remote celestial objects, radio-telescopes able to vie with optical telescopes of a large aperture would be required to be at least as large as the moon!

Man-made electrical interference also tends to plague the radioastronomer. By international agreement most nations refrain from transmitting commercial and other broadcasts on frequencies carrying natural cosmic signals. Unfortunately this is only part of the problem. Artificial static from various types of electrical gear can be extremely troublesome. This in a way is analogous to the city lights and smoke hazes that so often interfere with optical astronomy. In order to eliminate as much of this man-made static as possible, impulses from radio-telescopes are generally amplified and then fed into computers whose function is to sort out the desired space sounds from the human ones.

Radio-telescopes represent great feats of engineering. One of the largest is the famous 250-foot dish at Jodrell Bank, near Manchester. Despite its vast dimensions it is able to follow celestial radio objects with great precision even when subjected to gale-force winds and heavy rains. The reflector itself weighs 800 tons and must retain its exact shape in spite of all the stresses and strains it is called upon to endure. Such great engineering problems are not always overcome. A few years ago a mammoth 600-foot dish was begun by the United States Navy at Green Bank, West Virginia. This ran into such technical difficulties that its construction was ultimately abandoned.

The parabolic dish is not the only form of aerial used by radioastronomers, although it is certainly the one most familiar to the layman. Sometimes a form known as a dipole array is used. This is much less impressive and is essentially a number of small, separate, identical antennas that feed their combined individual signals into a common radio receiver.

A dish antenna can be tuned to a number of frequencies or wavelengths, whereas a dipole array will receive only a narrow band of frequencies dictated for the most part by the lengths of the various metal rods and the distance separating them. A dipole array is both cheaper and easier to erect but its use does necessitate a greater area of land. The dipole array of the Mullard Radio Observatory at Cambridge, for example, extends for half a mile and covers two acres. The number of individual dipoles in such a system can run into thousands. In some systems it is possible to rotate the dipoles so that they may point in more than one direction, but in most systems they are immobile and can scan only the part of the sky to which our revolving, orbiting planet directs them. This is considerably less of a limitation than might be thought. During the course of a single year, a fair proportion of the sky passes directly over every point on the surface of Earth. Moreover, a radio-telescope is restricted neither by cloud nor by daylight so that the effective sky 'coverage' is reasonably extensive.

It would be invidious to attempt a comparison between the two respective forms of radio-telescopes. Both have distinct uses and should really be regarded as complementary. The rotatable dish – or steerable paraboloid altazimuth radio reflector, to give it its proper title – is used primarily to locate and examine precise radio sources, whereas the dipole array is intended for comprehensive surveys or sweeps of the sky.

The fixed type of radio-telescope has a number of variants, some of which are extremely elaborate. A notable example is the helical antenna at Ohio State University, which is essentially a complex of spiral antennas mounted on a frame of wire mesh.

The fixed-aerial type of radio-telescope can, however, be used in the accurate pinpointing of radio stars under certain circumstances. More than one aerial must be employed, and such a system is known as a radio interferometer. The two or more separate antennas are regarded as *parts* of *one* vast aerial system. (This might seem an absurdly easy way to avoid the expense and effort involved in erecting a single, large system but economy has to be paid for in other ways.) The pattern obtained from two or more parts of an aerial system must of course be different from that which a single, giant instrument would provide. The pattern is in fact considerably more complex and from it the precise positions of the radio stars have to be calculated.

An interferometer generally comprises a fixed aerial, which may be as long as 3,000 feet, working in conjunction with one or more smaller, movable aerials, which have an approximate length of about 100 feet. Such a system corresponds to a single, large aerial covering several acres.

An arrangement of this nature on a grand scale is possible if two steerable dish instruments, separated by a hundred miles or more, are operated in unison. This enables the actual diameters of radio sources to be carried out with extreme accuracy. The two instruments focus on the same radio source thereby acting as a single radio-telescope having a diameter equal to the distance between them. When radio waves arrive at a radio-telescope aerial they alternately reinforce and diminish themselves in a system of concentric rings. This is known as an interference pattern and may be likened visually to the effect produced when two stones are flung simultaneously into a pool of water. Two coupled radio-telescopes focused on the same radio source receive an identical interference pattern. Earlier we saw that the greater the diameter of the dish the greater is its ability to identify radio sources lying close together. An instrument having an effective diameter of 100 miles is clearly at a great advantage. Moreover, it enables a much higher degree of accuracy to be obtained in respect to the direction from which radio waves are being received.

Radioastronomers do not, of course, *listen* to the stars and distant galaxies in the accepted sense. True, celestial objects make definite sounds but these in fact impart little real information. The 'signals' emanating from a radio-telescope are instead made to actuate the ink needle of a recorder that inscribes on a roll of graph-type paper a visual rendering of their form. The accompanying diagrams give some idea of the type of graph that can result. Figure 12 is a typical record of the signal received from a radio star, whereas Figure 13 portrays the

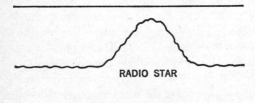

Figure 12

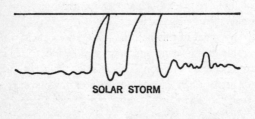

Figure 13

effect of a 'noise storm' on our own star, the sun. Both are examples of signals received by a radio-telescope of the steerable dish type. Those received from fixed-aerial instruments tend to be different, and an example of what might be received is shown in Figure 14a. In this instance the diurnal rotation of Earth has swept a pair of radio sources over the fixed-antenna system of the instrument. It will be seen that as the first source begins to come into line a small ripple shows itself. The signals are at

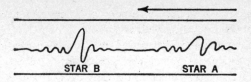

Figure 14a

their strongest when the source is directly in line, and consequently the amplitude of the recorded peaks are at their greatest. The aerial is then carried past the source, and so the signal diminishes. The process is repeated as the second and stronger sweeps into position.

A general and greatly simplified picture of the process can be seen in Figure 14b.

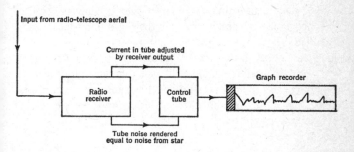

Figure 14b

Signals from a radio source are seen reaching the aerial dish of a radio-telescope. These are passed to the receiver, but here a difficulty arises. Any radio receiver from a normal domestic type to the finest communications model produces its own inherent noise. This bears no relation to the received signals flowing into the receiver from the antenna but originates within the instrument itself. Good design and special compensatory circuits can reduce this considerably, but it is virtually impossible to eliminate it entirely. Receivers for use in radio-telescopes are so designed that this noise is kept to a minimum. Such noise that remains, however, is used to estimate the strength of the cosmic signals. The inherent noise is controlled in such a way that its strength is kept equal to that of the received radio noise arriving from the aerial. The noise is thus transformed into an exact replica of the cosmic signals that when duly amplified are used to actuate the recorder needle and so produce a graphical interpretation.

We are still very much in the early days of radioastronomy, and, although the science is barely four decades old, the strides this *enfant terrible* has made can only be described as startling. It is both ironical and tragic that Karl Jansky, who first drew the attention of his fellow men to the natural radio emanations originating in the depths of space, did not live to see the realization of his dreams. All through life he had waged a brave but losing battle against ill-health, and in 1950 he died at the early age of forty-four. But already Karl Jansky had become a legend in his own lifetime. His epitaph is secure – he was the founder of radioastronomy.

But what of our main theme? Great galaxies, glowing nebulae, and fiery suns are speaking to us in their own unmistakable and inimitable language. What we must now seek to ascertain is whether or not all the natural babble and chatter of the great cosmos is concealing a still small voice that tells not of distant, wheeling galaxies or of remote exploding suns, but of the hopes, fears, and aspirations of other intelligent creatures on worlds we can never hope to see.

Part Two

8 Searching the Void

It is easy to become so fascinated by the idea of interstellar communication that the difficulties are temporarily forgotten. When, however, a move is made towards realization the initial feelings are a combination of helplessness and frustration. Just where can a start be made on a project that is so complex and so far-reaching in its implications?

It is apparent that *optical* telescopes can never be a means of searching the universe for other civilizations. The greatest present-day instruments are unable even to reveal the stellar planets on which alien civilizations, if they exist, must be located. Were such telescopes located on the moon or some other relatively airless body such a revelation might come within the bounds of feasibility. Even then it is extremely doubtful if precise information regarding these incredibly distant worlds would be forthcoming. Indeed we know surprisingly little concerning our own sister world of Pluto even though astronomers have been aware of its existence since 1930. Planets are, relatively speaking, very small celestial objects that shine only as a result of reflected light, light that is provided exclusively by the parent star. If this, then, is a measure of the difficulties within our *own* solar system, how much greater must be the problem in relation to suns and stellar systems several light years removed?

We must therefore turn to the radio-telescope, hoping that this instrument, by virtue of its ability to detect natural radio emanations, will one day indicate the existence of intelligent alien signals.

Recent research has already shown that large radio-telescopes should be capable of detecting high-power radar signals over a distance of 8·7 light years. This range is sufficient to include a number of appropriate stars. Thus we are already in possession of equipment fairly adequate for our purpose.

The use of a parabolic reflector as a receiving antenna gives us the type of instrument most commonly associated with the radio-telescope. There is a rather handy little rule or relationship that tells us that if the antenna diameter (expressed in feet) is divided by ten the result represents the distance (in light years) at which alien high-power transmitters, similar to our own, can be detected.

The accompanying table gives an idea of the position today with respect to a number of the largest existing parabolic radio-telescopes, both steerable and fixed.

Instrument	Diameter (Feet)	Range (Light Years)
Jodrell Bank, Cheshire (steerable)	250	25·0
Puerto Rico (fixed)	1,000	100·0
Green Bank, West Virginia (steerable)	85	8·5
Naval Research Laboratory, Washington, D.C. (steerable)	50	5·0

TABLE 6

A few years ago construction was begun on a 600-foot diameter dish in West Virginia at a place called Sugar Grove, 30 miles north-east of Green Bank. This radio-telescope would have given a theoretical range of 60 light years, and since it was to be of the steerable variety it might have proved an invaluable asset to the interstellar communication armoury. Unfortunately the project failed because of engineering difficulties that proved impossible to overcome. If such an instrument is to have satisfactory accuracy the parabolic reflector must be true to within one-tenth of the wavelength being received. Such a specification demands the almost complete absence of any sag, an almost impossible condition to satisfy in the circumstances. Only by considerable bracing and support could it have been met, a modification which would have resulted in a steerable dish weighing approximately 36,000 tons! It is rather obvious why the project was abandoned!

It is estimated that the instruments presently in existence bring approximately 10,000 stars within radio range of this planet. It must be plainly understood, however, that 10,000 stars do not mean 10,000 planetary systems. Nevertheless, present cosmological belief is that stars are rarely formed as single entities. Either two or more separate suns evolve from the primordial nebula, in which case the end-products are binary or multiple star systems, or a single star is formed in conjunction with a number of much smaller non-stellar bodies, that is, planets.

Planets of course do not automatically connote the existence of life any more than stars imply planetary systems. In the nine planets of the solar system it is fairly certain that apart from Earth only Mars is at all likely to nurture a form of life and even this could only be of a very low order (e.g., a form of lichen). We are, however, looking for life which is not just biologically advanced. The object of our search is life that has had time and has been able to achieve a highly sophisticated technology.

There is little doubt that worlds containing life, primitive or advanced, cannot possibly be as plentiful as planets themselves. The main drawback lies in the simple fact that for its initiation and development life demands a rather narrow and constricting set of physical conditions. Biologists believe that, given a planet sufficiently Earth-like, there is no valid reason why life-initiating reactions should not take place since the environment will be appropriate. Unfortunately, though we speak freely of life-initiating reactions and mechanisms we do not know their real nature.

Although in our search for life we look primarily for Earth-like conditions it must be accepted that a high proportion of planets are probably not Earth-like in the least. Out of the eight other known major planets in the solar system, two at most can be included in this category – Mars and Venus. Compared with Jupiter and Saturn these two worlds *are* probably Earth-like; compared to Earth itself they certainly are not!

In this context, Earth-like refers to a planet that

1. has a stable central star akin to the sun in respect of temperature;
2. lies within a zone receiving the same amount of light and radiation as does the Venus-to-Mars zone within the solar system;

3. is neither too large nor too small;
4. has a rate of axial rotation similar to that of Earth;
5. contains solid land as well as ocean;
6. has an atmosphere of nitrogen and oxygen akin to that of Earth.

It can be argued that by virtue of adaptation life might have evolved on planets that do not conform to some of these conditions. This may well be so, and as long as we do not carry the argument to absurd extremes we are entitled to envisage life on worlds fairly different from our own.

We have seen in Chapter 3 that there is a considerable disparity in the ages of stars. Therefore, the ages of extrasolar planets must also differ widely. Some will be new and evolving worlds, others will be very old indeed. Between these extremes there must exist a wide range of ages. It is from worlds that are either contemporaries of our own or older that we must anticipate intelligent radio signals. Thus our search should be directed towards stars whose age is equivalent to or somewhat greater than that of the sun.

Any alien civilization seeking to establish contact with a cosmic neighbour will presumably be faced with problems very similar to those that confront our own radioastronomers. The question of frequency is obviously of fundamental importance. A channel in which the radio waves suffer a high degree of attenuation in their passage through interstellar space is of little practical use. This applies also to a channel in which these waves become hopelessly weakened or lost in passing through the atmosphere of the 'destination' planet. If, of course, the receiving world has space stations lying above its atmosphere the latter difficulty is resolved. Unfortunately the species on the 'transmitting' planet will almost certainly be unaware of the existence of these stations.

Radio frequencies lower than 1 mc/s and those above 30,000 mc/s are regarded as the most likely to suffer absorption in passing through planetary atmospheres. Those in the near light and gamma regions of the spectrum, though perhaps theoretically acceptable, would entail the use of extremely high-power equipment as well as the employment of complex and highly specialized techniques. We are thus left with that region of the electromagnetic spectrum lying between 1 and 10,000 megacycles per second.

Before proceeding further, two other factors must be taken into account. The first factor is that a race bent on radiating signals to another planetary system must consider the inevitable background emission of its own sun. The second factor is that of background noise originating in the galaxy itself. A terrestrial aerial directed towards a particular star system will also receive galactic background coming from *behind* the star.

Calculations based on the natural emissions of a star similar to the sun, allowing for an appropriate degree of galactic emission, indicate that for a source signal of 10 light years distant, *minimum* background noise will be found at 10,000 megacycles per second. This is computed on the basis of an antenna having a diameter of 100 metres.

This does not mean that least background noise will be encountered only at 10,000 megacycles per second, or that such noise will exist at a high level at all other frequencies. Rather we must regard 10,000 megacycles per second as the central point of a broad-frequency band in which noise due to stellar and galactic emission will be at a low level. This means that the major portion of the region already specified as generally suitable can also be regarded as satisfactory from the point of view of noise level, that is, 1,000 to 10,000 megacycles per second.

So far so good, but the realities of the situation seem as formidable as ever. We must recognize clearly what our aim is. It is to search for signals that, if they exist at all, will be very weak and of an indeterminate frequency. Noise level, no matter how low, must still be regarded as significant if the signal sought for is very faint. Moreover, a really close search of the entire 1 to 10,000 megacycles per second band constitutes a not inconsiderable undertaking in terms of both time and effort, if indeed such a search is practicable at all. It is obviously going to be a distinct advantage to reduce in extent the region of the electromagnetic spectrum we must search. Is this possible?

But for a happily fortuitous circumstance the answer might well be no. It happens, however, that nature has provided us with a possible key. At a wavelength of 21 centimetres, corresponding to a frequency of 1,420 megacycles per second, we find the emission noise of the great clouds of hydrogen, which are so essential a feature of our galaxy. To every technologically advanced civilization this must be a well-known fact, the significance of which can hardly fail to be appreciated.

For a moment or two let us put ourselves in the position of an alien and advanced race. These beings might very well reason along the following lines as they look out into their night sky and see the bright yellow star that is our sun. 'There,' they may say, 'is a star that probably possesses planets.' (They too will have well-founded cosmological reasons for believing this to be so.) 'Some of these planets may be inhabited by intelligent creatures. If so, these creatures will probably appreciate the significance of a frequency channel of 1,420 megacycles per second since it is that of the most common cosmic emission. It is the first that should have been picked up by their radioastronomers, and if it has it will be the subject of much close examination. Here then is the frequency on which intelligence can best be transmitted, for it is the one most likely to be listened to.'

In the first instance at least, *our* search then should be conducted on and in the immediate vicinity of this frequency. Though this represents a sensible philosophy there looms a possible objection. A few paragraphs back we stressed the point of background noise and how it might well obliterate or conceal a weak intelligent signal. Surely the emission sound of these hydrogen clouds will do just this very thing?

Fortunately the galactic background due to these clouds is by no means uniform. In those directions near to the plane of the Milky Way it is some forty times greater than in others. Thus it will obviously be a sensible policy to examine first the suitable stars lying well away from the galactic plane. In some respects this is unfortunate since Alpha Centauri, 70 Opiuchi, and 61 Cygni are all viewed from Earth against the background of the Milky Way. The last two are both thought to possess planetary companions while Alpha Centauri, though a multiple system, has a solar-type component and is, of course, close to us, astronomically speaking. However, several well-known stars of particular interest in this context lie well clear of the plane of the Milky Way.

It has been calculated that in order to produce in our receivers a signal equal in strength to the galactic background, a transmitter on the planet of a star 10 light years distant would, were it using a parabolic reflector 80 metres in diameter, require a degree of power that would perhaps not be beyond *our* capabilities but would certainly tax them to the fullest. The use of a radio-telescope having an antenna diameter of 200 metres would, however, call for a markedly lower power output and one that would certainly come within our capabilities. Unfortunately the disastrous experience at Sugar Grove renders this of only academic interest at the present! Nevertheless, we are basing our premise largely on beings to whom we are technological inferiors, and consequently the question of power capability is unlikely to be a particularly valid one.

It is essential that we reason in a similar vein when we come to consider the number of radio beams that an alien race might set up. Both Guiseppe Cocconi and Philip Morrison, renowned specialists in the field, regard 100 such beams as not impossible to a society more advanced than our own. They conclude that we can thereby hope to detect a beam directed towards the solar system from any appropriate star within a few tens of light years.

Our deliberations then lead us to the conclusion that our search should be conducted on or around a frequency of 1,420 megacycles per second and that our antennas should be directed towards appropriate stars (preferably of a G-type main sequence) lying *outside* the galactic plane. Having reached this conclusion there is yet another factor that we must consider.

When a body emitting radiation moves either closer to or farther from us, an *apparent* change or shift takes place in the frequency of that radiation. This is the celebrated Doppler Shift, the effects of which are not uncommon. A classic example is the change in the note of an express train whistle which occurs when it rushes past a stationary observer. Optical astronomers are, of course, well aware of this. Because of this phenomenon the recession of the galaxies was first revealed. These great galaxies when examined spectroscopically show a definite shift towards the red end of the spectrum – a clear indication that they are in headlong retreat.

A measure of the Doppler Shift can sometimes be anticipated in the frequency of signals transmitted by beings on extrasolar planets. Such planets will orbit their parent stars just as Earth and all the worlds of the solar system sweep around the sun. In many instances these sources of artificial radiation will alternately recede and approach with respect to Earth. (See Figure 15a and Figure 15b.)

In the diagram in Figure 15a the orbital plane of the alien planet is such that the planet moves infinitesimally towards Earth and also recedes from it to the same extent. Figure 15b illustrates an entirely different state of affairs, for in this instance the plane of the planet's orbit is at right angles to the plane in Figure 15a. In this case the planet does not recede or approach relative to Earth, and consequently we would not anticipate any Doppler Shift.

To what extent would this alter the frequency of the radiated signal? Cocconi and Morrison have suggested that it would cause a fluctuation of about 300 cycles per second. Although this represents a very small shift it must still be taken into account.

While dealing with the subject of frequency we should add that the search 'area' could probably be extended beyond the

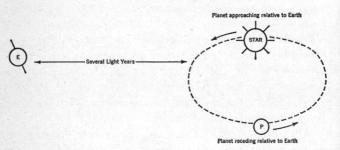

Figure 15a

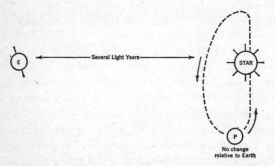

Figure 15b

upper limit of 10,000 megacycles per second were reception to take place on satellites or space stations located well beyond the earth's atmosphere. Since Earth is known to possess an absorbing effect on radiation it seems not at all unlikely that eventually reception by satellites will be tried out. Above 10,000 megacycles per second radiation noise level from the terrestrial atmosphere rises very steeply. The obvious snag about reception above the atmosphere is, of course, the fact that so far it is not possible to place huge parabolic-type radio-telescopes in orbit. However, the pace of technology should render this a feasible proposition in the not too distant future.

Thus another advantage of establishing bases on the moon becomes apparent. Here is a world virtually devoid of atmosphere that interferes with or absorbs electromagnetic radiation. Moreover the much lower lunar gravity would greatly facilitate the erection of giant radio-telescopes.

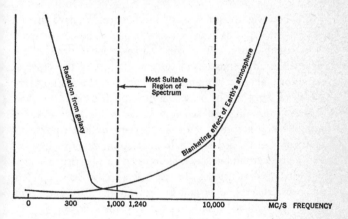

Figure 16

The graph in Figure 16 outlines the position with regard to frequency. We know that from 1 to 1,000 megacycles per second galactic radiation is at a high level and would therefore smother faint signals. Beyond 10,000 megacycles per second, Earth's atmosphere exerts a similar blanketing effect. Between 1,000 and 10,000 megacycles per second lies the suitable band of frequencies that includes also the especially desirable 1,420 megacycles per second.

Now, of course, another important point comes to the forefront. Would the signals from the world of another star have the necessary range? Would they in fact have the strength to reach us? The position in this respect is regarded as very favourable. The United States probes to the environs of Venus and Mars have already shown the transmission of intelligence over vast distances to be a perfectly feasible proposition. Admittedly the distances involved were only of the order of several millions of miles and therefore far removed from the awesome light years that separate us from the stars. However, it should be borne in mind that what has already been accomplished by these probes was done with miniaturized radio transmitters of low-power output. Even now the feasibility of establishing radio links over distances in excess of 10 light years is accepted with the provision that very large antennas and ultrasensitive receivers be used. No insurmountable difficulty should be encountered in respect of range.

More, however, must be said of interference. We have already seen that the types of radiation most likely to be encountered can be attributed to two main sources. One source is the unavoidable radio emission of the star itself, and the other is equally inevitable background noise from the galaxy. To overcome the former would present no insuperable problem. The transmitter beaming the signal towards the star with which communication is sought must simply have sufficient power to make itself heard above the natural radio emission of its own star. The background galaxy noise represents a barrier of somewhat greater dimensions. But as has already been pointed out, such background is not uniform in intensity and from regions well removed from the galactic plane it is appreciably less.

The next point for consideration concerns the characteristics of received interstellar signals. These are of paramount importance, and it is essential that we know precisely what we are looking for.

The narrower the bandwidth of a transmission the greater its effective range. This is one of the fundamentals of radio communication. Interstellar contacts involve tremendous distance and therefore it is reasonable to expect such transmission to be of narrow bandwidth.

The term bandwidth should perhaps be explained at this point. In this respect it might be an advantage to consider the transmission from a conventional broadcasting station. Suppose, for example, that the station transmits on a frequency of 200 kilocycles per second, equivalent to a wavelength of 1,500 metres. If we tune a radio receiver to 1,500 metres we will hear the station, assuming that the station is then transmitting and that the receiver dial is accurately calibrated. This does not mean that on 1,499 metres or 1,501 metres we will hear nothing. Quite the contrary, in fact, for as we turn the tuning control the first sounds from the station will be heard around 1,450 metres but they will be faint and probably distorted. In the immediate vicinity of 1,500 metres the signal will be at its loudest and clearest. As we continue past this vicinity the signal becomes progressively fainter and more distorted until at around 1,550 metres it disappears entirely. We can say, therefore, that this particular transmission has a bandwidth of approximately 100 metres. An amplitude-modulated signal of this sort – in other words, a normal broadcast – has what are known as side bands and these, as we have just seen, extend to some distance on either side of the fundamental frequency or wavelength. Had it been a straight Morse transmission, that is, an unmodulated, punctuated carrier wave, these side bands would have been virtually non-existent. In other words the bandwidth would have been very narrow indeed and we would have heard the signal on 1,500 metres and to a few metres on either side of it.

The slight shift in frequency brought about by virtue of the Doppler effect, though it might prove a minor nuisance in some respects, could be of appreciable value in helping us distinguish intelligent signals from natural cosmic background and terrestrial interference. Changes in signal strength, also a handicap in some ways, could prove a further distinguishing feature and might be due either to a form of coding or to the revolution of the transmitting planet on its axis, or both.

Fundamentally the signal should come from the direction of a particular star. This star should be one of suitable type and relatively near by stellar standards. A signal emanating from a point in space where apparently no close star exists must be suspect unless other very good grounds exist for believing the radiation to be of artificial origin. A signal showing a clear pattern, such as a system of pulses and gaps or dots and dashes, would be of immediate interest even though the signal originated from a point in the heavens apparently devoid of any star. We must bear in mind the strong possibility of an optical telescope revealing a star at that point. Such a star need not necessarily be very remote. It could be one of relatively low heat and light output lying 'close' to the solar system. A typical instance is provided by the star that lies closest of all to our sun but nevertheless cannot be seen with the unaided eye. This is the well-known Proxima Centauri, which is only 4·2 light years distant from the solar system and whose light output amounts to only 1/10,000 that of the sun. This star should not be confused with its well-known companion, Alpha Centauri, which is a bright, first-magnitude solar-type star lying 4·3 light years from us. We are not inferring that Proxima Centauri is necessarily a star from whose immediate environs we can expect intelligent signals – indeed this is rather unlikely. We merely seek to show that it is possible for a signal to emanate from a point in the heavens relatively close to ourselves even though no star may be visible to the unaided eye.

Recapping further, we can say that artificial radio emanations from the stars would be distinguishable from the natural variety in the following ways:

1. A radio emission of artificial origin might show short-period Doppler Shifts due both to axial rotation of the planet and to its orbital movement around the parent star.

2. Artificial radio transmissions could be expected to have a narrow bandwidth compared to the fairly wide ones associated with natural emissions.

3. The continuous wave generated by an alien transmitter might vary in strength or be punctuated by a system of coding.

It is fascinating to consider where interstellar communication might eventually lead us. It is possible that years from now our descendants will listen to the weird cadences of an alien symphony playing on a world of Epsilon Indi, hear with helpless pity the plea for help from the dying planet of Tau Ceti, be appalled by the cacophonic rantings of a galactic dictator seeking to extend his hegemony over yet another world of Epsilon Eridani?

Star cloud in Sagittarius, source of first discovered radio emanations from space. (*Courtesy of the Royal Astronomical Society, and Lick Observatory, California*)

Above Great Spiral Galaxy in Andromeda showing also satellite galaxies NGC 205 and 221. (*Courtesy of Mt Wilson and Palomar Observatories, California*)

Opposite Cygnus 'A' radio source. (*Courtesy of Mt Wilson and Palomar Observatories, California*)

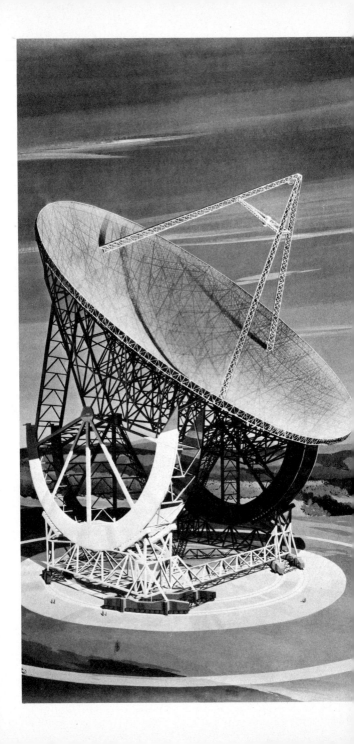

Above World's largest radio-telescope, Muenstereifel, Germany. (*Courtesy of United Press International, Inc., New York*)

Opposite Artist's conception of 600-foot radio-telescope, Sugar Grove, West Virginia. (*Courtesy of United States Navy and American Museum of Natural History, New York*)

Above Antenna 'mirror' base of world's largest radio-telescope, Muenstereifel, Germany. (*Courtesy of United Press International, Inc., New York*)

Opposite above The difference in transit times between interplanetary travel and interstellar travel. Artist: Nick Cudworth. (*Reproduced from ICI Magazine by permission of ICI Ltd.*)

Opposite below The principle of time-dilation interstellar travel. Artist: Nick Cudworth. (*Reproduced from ICI Magazine by permission of ICI Ltd.*)

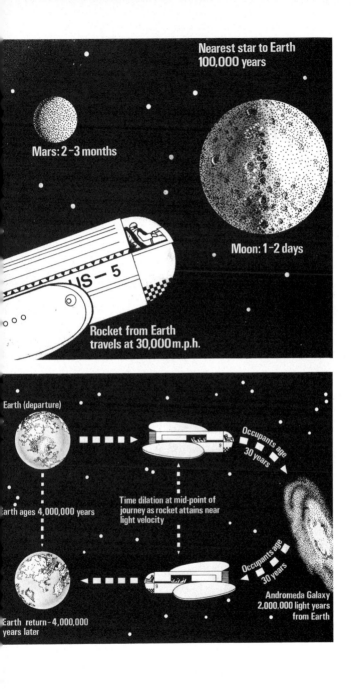

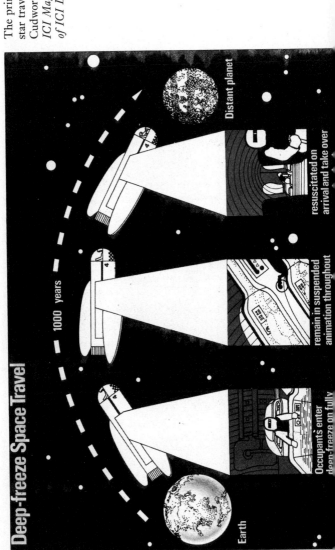

The principle of generation star travel. Artist: Nick Cudworth. (Reproduced from *ICI Magazine* by permission of ICI Ltd.)

It is important to remember that the signals we pick up would not just be messages from space. They would also be messages out of the past. The signal that whispers so softly in the aerial of a terrestrial receiver, that stirs so infinitesimal a current, is weak from its terrible journey through the black depths of eternal night. It might tell of events years past, of great civilizations from which the glory has long since departed. It could be a tale of great joy – or infinite sadness!

The time aspect must count for considerably less in the case of relatively close planetary systems. The messages we might receive from Tau Ceti could tell of events that took place there a mere eleven years ago, and signals from Alpha Centauri could speak of happenings only four years past. It is only on the day when we build an electronic bridge capable of spanning the greatest galactic gulfs that the poignancy of receiving signals from beings and civilizations long dead arises. Perhaps this particular aspect will never be of more than academic interest. Despite this, its implications remain intriguing. Had the inhabitants of this planet of ours two and a quarter million years ago been able to beam messages into the great deeps of space the worlds on the fringes of the Great Andromeda Nebula would only now be receiving the latest news and views of the dinosaur family!

Science-fiction writers over the years have frequently made us aware of the strange implications and peculiar pathos that might result from the development of interstellar travel. Now we begin to realize that interstellar communication has its share of these things, too. From worlds we can never hope to see, our own world, if it has the ears to hear, the patience to listen, the mind to understand and truly comprehend, may yet secure the pass-key to a new and enlightened future.

9 Which of the Host?

WE have decided in which general directions we should direct our radio-telescopes. Now we have to be more specific and think in terms of individual stars.

To some extent our thinking must be on similar lines to that of the first interstellar space expedition, which will look for a star whose age and type renders habitable planets likely. We must also look for such a star. In our case, however, we require planets not only habitable but *inhabited*, and these planets must be of sufficient age for their peoples to have gained a high degree of technological maturity.

Distance is of fundamental importance. So far as interstellar expeditions are concerned this is too obvious to require elaboration. When we direct our radio-telescopes towards stars in the hope of intercepting messages, distance must figure largely in our calculations too. The nearer the star world sending signals the stronger these should be. The signals, therefore, will be easier to pick up and identify. Moreover, if we are to reply and so establish proper contact it is highly desirable to reduce the time factor as much as possible. This is important, remembering that contact between ourselves and even the nearest star must involve a minimum period between 'question' and 'answer' of almost nine years.

For these reasons we will consider only our closest stellar neighbours, stars up to 5 parsecs (16·3 light years) from the sun. This involves some 40 stars, nine of which are double and two triple. A few of these stars are believed to have planetary companions.

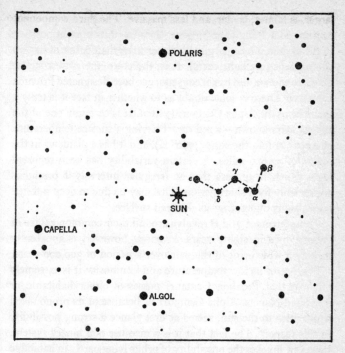

Figure 17. The sun from another star

How a race of beings on a hypothetical planet of the binary star system Alpha Centauri, bent on contacting solar planets, would see the sun – merely as a very bright yellow star close to the familiar W of the well-known constellation Cassiopeia.

Alpha Centauri

This triple system must figure largely in our calculations since it is the closest of all the sun's stellar neighbours. Because it lies near to the Southern Cross it is never seen from the Northern Hemisphere. Being a first-magnitude yellow star it is easily identifiable and is in fact the third brightest star in the sky. Its distance from us is 4·3 light years (just over 25 million, million miles). The two main components of the system are known as Alpha Centauri A and Alpha Centauri B. The former is the brighter and is similar to our sun in respect of size, mass, temperature, and luminosity. Alpha Centauri B, though slightly larger, is fainter, cooler, and less massive. The third component remains something of an enigma. Because of its present position in the system it lies marginally closer to us than either of its two companions (4·2 light years). It is, therefore, the closest stellar object to the sun and has in consequence been designated Proxima Centauri. There is some doubt as to whether in fact it is truly a member of the Alpha Centauri system. It is certainly one of the coolest stars known – a red dwarf having a surface temperature of 2,000° K (cf., the sun, 5,700° K), and it has a diameter in the region of 52,000 miles. A certain variability has been reported from which it appears that at irregular intervals it brightens appreciably for a short period. This may be due to deep internal convulsions puncturing its dull red surface.

Components A and B revolve around their common centre of gravity once in eighty years. Proxima, however, is apparently making a wide orbit of the pair over a period of 300,000 years.

Because of its low temperature and luminosity it is extremely unlikely that Proxima Centauri possesses either habitable or inhabited planets. Alpha Centauri A, because of its pronounced similarities to the sun, seems at first glance a strong possibility in this respect. The fact that it is a member of a binary system, however, invokes the possibility of orbits lying partly in habitable and partly in non-habitable zones.

Barnard's Star

This red dwarf, so similar in nearly all respects to Proxima Centauri, is at 6 light years the second closest star to the sun. The existence of a planetary companion was fairly definitely confirmed in 1963, but in view of the star's low luminosity it is unlikely that any planets it happens to possess could be inhabited.

Sirius

Sirius, alias the 'Dog Star' because it is the principal star in the constellation of Canis Major, the Great Dog, lies 8·7 light years from the sun. It is therefore about twice as far from us as Alpha Centauri. Sirius, twice as massive as the sun and with double its diameter, is a young, white, very hot star having a temperature in the region of 11,000° K. In the circumstances it is hardly surprising that it is the brightest star in our skies. A more striking impression may be gained if we imagine Sirius as lying at the same distance from earth as the sun. If it did, our planet's surface would lie barren and scorched under a torrent of heat and light forty times greater than that which it presently experiences. Because of this profligate expenditure of energy Sirius has a much reduced life expectancy. Such a young star is most unlikely to possess planets. Consequently there seems little point in turning our electronic ears towards it.

Beyond the 8·7 light years of Sirius we find a number of stars that can fairly and safely be disregarded as centres of intelligent life. These are all red dwarf stars, similar to the 'cool', dull Proxima Centauri.

A little farther out but still within the limit of 5 parsecs we come to a couple of K-type stars. Both stars are orange-yellow in colour, hotter and larger than the sun, and extremely interesting from our point of view. The first of these is Epsilon Eridani.

Epsilon Eridani

To the naked eye this is an undistinguished star in the long straggling constellation of Eridanus, the River. Its potential as a probable centre of advanced life has been recognized for some time, and it was one of the two stars to which the 85-foot radio-telescope at Green Bank, West Virginia, was directed during the Project Ozma experiment of 1960. This star lies 10·5 light years from the solar system and will almost certainly be the subject of future attention in the realms of interstellar communication.

61 Cygni

This is the second of the two Class K stars. Like Epsilon Eridani it too is fairly undistinguished, lying 11 light years from the sun. As its title implies it is situated in the lovely summer constellation of Cygnus, the Swan (known also as the Northern Cross).

Known to be a binary system, 61 Cygni is also believed to contain a dark non-stellar object, which as a planet would be too massive to sustain life. However, it is not unreasonable to assume that where a large planet exists so also may smaller, more appropriate ones. 61 Cygni must therefore, in our context, remain a star of considerable interest.

Procyon

This brilliant yellow-white star lies 10·4 light years from Earth and the sun. It is one of the brightest stars in our skies and one of the six first-magnitude stars surrounding the magnificent winter constellation of Orion, the Hunter. Procyon has a surface temperature of 7,000° K and like Sirius possesses a small unseen companion.

Next in order of increasing distance we come to a number of rather small and relatively cool stars. Four of these stars are of special interest to us.

Epsilon Indi

This is a single star 11·4 light years remote. It is of K-type, orange-red in colour, and can only be seen from the Southern Hemisphere. Even though it is rather on the 'cool' side, Epsilon Indi is reasonably favourable from our point of view.

Tau Ceti

Like Epsilon Indi, this too is a single star but of a solar type. Tau Ceti is probably one of the most favourable stars so far as interstellar communication is concerned. With a surface temperature of 4,500° K, it is appreciably cooler than the sun. This star is a rather insignificant object lying in the constellation Cetus, the Whale, and is 11·6 light years distant.

Omicron Eridani

In this instance we have a triple system of suns lying 16·2 light years from us. The most luminous of the three components is slightly cooler and smaller than the sun. The second, red in colour, has a mass and volume about one-fifth that of the sun. The remaining member of the trio is a white dwarf having something of our sun's mass yet only 2 per cent of its size. The red component and the white dwarf circle each other in approximately 250 years, making a wide orbit of the bright member as they do so. Because of its multiple nature this system is not regarded too favourably planet-wise.

Altair

This bright star in the constellation Aquila, the Eagle, lies 16 light years distant and is therefore just within the boundary of 5 parsecs (16·3 light years), the acceptable distance set earlier. Altair is a hot, white star and has no companion. This star has an extremely high rate of rotation in comparison with most other stars – 160 miles per second at its equator (cf., 1·3 miles/second for the sun). As a consequence it may not be spherical and is probably flattened at the poles with a bulge in its equatorial regions. It seems improbable that Altair, a young star, will be found to have a planetary retinue. In due course belts of matter may break away from its violently spinning girdle to form a series of worlds. Under these circumstances Altair can be of little interest to us.

We should now be in a position to draw up a list of stars appropriate to our purpose. From our present list, Sirius and Altair can immediately be eliminated since both are considered too young to have planets. Procyon, though suitable in some respects, is only just acceptable so far as age is concerned.

Double (and multiple) star systems have of necessity to be regarded with doubt. The components of such systems are thought to have been formed simultaneously. If this is the case, then, little material to condense into planets may have been left.

This, of course, presupposes that planets are born not from a central sun but from residual matter.

The components of binary systems are on the average about 20 astronomical units apart (1,860 million miles), which is roughly the distance between the sun and Uranus. In these circumstances it is difficult to conceive of a stable orbit *within* the system although one around *both* components seems reasonably acceptable.

Red dwarf stars, so far as planetary formation is concerned, remain something of an enigma. They are favourable in two respects – they rotate slowly and are of sufficient age. Unfortunately these criteria alone are insufficient. Because of the low mass of red dwarfs they may never have achieved a 'planet-spawning' state. Moreover, their low heat and light output is an inhibiting factor since the habitable zone for planets orbiting such suns would be restricted to narrow regions close to them. Indeed to be habitable a planet of a red dwarf would be required to orbit the parent star much more closely than Mercury does the sun (33 million miles).

Our choice is almost automatically narrowed to the stars of spectral group G (solar type) and the hotter stars of group K. The two main components of Alpha Centauri are G-type and K-type respectively, but the fact of a binary system has to be considered. The separation of the two is 23·4 astronomical units, a distance that is probably sufficient to permit the planetary orbit of either component. To date, perturbations in the Alpha Centauri system have not been reported. This of course merely implies that large, massive (and therefore unsuitable) planets are not present. Small and quite appropriate planets could still exist. It is held in some quarters that the Alpha Centauri system is considerably younger than the sun and that as a consequence life mature enough for our purpose will not yet have developed.

The only other G-type star within 5 parsecs of the sun is Tau Ceti. This being a single body is a favourable choice in all respects except distance. Among K-type stars, Epsilon Eridani and Epsilon Indi are of definite interest. Both are single, but Epsilon Eridani is the more likely possibility because it is warmer than Epsilon Indi. Any inner planets Epsilon Eridani possesses are probably habitable. As we shall see, Tau Ceti and

Epsilon Eridani were the two stars favoured in Project Ozma, the first attempt to intercept alien radio signals.

61 Cygni is a binary with both components being K stars. The probable presence of a dark non-stellar companion at once renders this system of interest. The mass of this superplanet works out at approximately sixteen times that of Jupiter. The gravity of such a world must be regarded as 'body crushing'. We might hope, however, that it possesses satellites of suitable gravity or that, better still, the twin suns of 61 Cygni also possess planets of much more modest proportions.

Tau Ceti and Epsilon Eridani are therefore the ideal first choices on virtually every count except that of distance. In respect to distance Alpha Centauri, even though it is a binary, has a clear lead and must be regarded as the next choice. 61 Cygni and Epsilon Indi follow. We must also mention 70 Opiuchi, which is just on the 5 parsec limit. It too is a binary with both components being of class K. 70 Opiuchi is strongly suspected to have a dark massive companion.

It should be added that Tau Ceti, Epsilon Eridani, and Epsilon Indi are well removed from the galactic plane. Alpha Centauri and 61 Cygni on the other hand have the disadvantage of the galaxy (and its interference) as a background.

10 Project Ozma

IN 1952 a young man graduated from Cornell University who had long been fascinated by the idea of other worlds and alien civilizations. At Cornell he had heard the celebrated astronomer Otto Struve lecture on the subject of slow-rotating stars and their probable connection with planetary systems. Following three years in the United States Navy, in which he served as an electronics officer, the young man proceeded to Harvard University where in due course he received a doctorate in astronomy. The name of this youthful scientist was Frank Drake.

While at Harvard, Drake became increasingly interested in the fascinating and rapidly advancing science of radioastronomy. In conjunction with Harold Ewan, one of the discoverers of the 21-centimetre radio emissions, he studied the possibility of using these emissions as a means of solving certain astronomical problems. In 1958 he secured an appointment to the newly-created National Radio Astronomy Observatory at Green Bank, West Virginia. Inevitably his thoughts turned again towards the subject that for so many years had intrigued him. Now, however, new and distinctly more exciting possibilities were stirring in his mind.

On arrival at Green Bank, Drake faced a scene calculated to depress the spirits of the most enthusiastic person. The National Radio Astronomy Observatory, despite its grandiose title, was still more a name than a reality. No permanent buildings had yet been erected, the observatory offices were in an old farmhouse, and last but by no means least, the entire site was a sea of snow and mud over which hung a sullen sky of winter grey. An environment further removed from the concept of advanced cosmic civilizations would have been hard to imagine. Frank Drake, however, was not the sort of man to be put off by adversities of this nature. In his mind he saw the finished observatory and felt something of the immense potential it would soon possess. Indeed a certain aura seemed even then to surround the area. Here in a quiet West Virginian valley was rising a listening-post to the stars.

The site for the observatory had been selected with considerable care and after much thought because the requirements of a radioastronomy observatory are both exacting and unique. The main essential, freedom from extraneous man-made radio signals and electrical interference, meant that the site must be well removed from cities, towns, and all other population centres. Green Bank lay in a valley screened to some extent from unwanted radio signals by nearby mountains. To improve its prospects further a law had been passed forbidding entry to all organizations or individuals using equipment in any way likely to produce undesirable radio emissions. In addition the Federal Communications Commission had established an area measuring 100 by 120 miles in which the use of all licensed radio transmitters was barred. Even aircraft were requested to avoid the area as far as possible. Aircraft, apart from radio interference, could with their metal bodies easily give rise to spurious reflection effects. Indeed, they were seen as the greatest potential source of trouble, for the desirable 21-centimetre channel lay in the middle of a frequency band already allocated to aviation. Thus side bands or harmonics of an aircraft's transmission might easily interfere with the designated channel even though it was not actually transmitting on that frequency. As we shall shortly see, such warnings proved prophetic.

The National Radio Astronomy Observatory was essentially a United States Government project since a scheme of this magnitude was felt to be beyond the resources of private organizations alone. Responsibility for establishing the observatory was, however, left to a consortium of nine universities known as Associated Universities. The head of this group, which had already been responsible for setting up the Brookhaven National Laboratory in Suffolk County, New York, was Lloyd Berkner, a noted radiophysicist and former member of Admiral Richard E. Byrd's expedition to the Antarctic. Berkner, well known as a man to whom unorthodox methods and novel schemes were second nature, had also been responsible for suggesting the International Geophysical Year of 1957–1958.

It was Berkner who acted as director of the new observatory, a state of affairs that augured well for certain ideas then rapidly taking shape in the fertile brain of Frank Drake. When Drake approached Berkner on the subject of his thoughts he was invited to present his case officially. Realizing that the tide was running in his favour Drake was at pains to emphasize the immense possibilities of a programme aimed at the detection of intelligent signals from space. He stressed how for centuries mankind had speculated aimlessly on the subject of extraterrestrial life and how it seemed the question might for ever remain unanswered. At last there was a chance that dramatic changes might be wrought. Drake added further emphasis to his argument by drawing attention to the extremely high degree of sensitivity possessed by contemporary radio-telescopes, and in conclusion he pointed to the economic viability of a programme that could for the most part be carried out with equipment already in existence or soon to be available.

Berkner was impressed not only by Drake's proposals but by the zest and enthusiasm of the young scientist. In the spring of 1959 he authorized Drake to go ahead. Drake at once decided on the name 'Ozma' for his pet project. Named after the mythical land of Oz, a remote, inaccessible region inhabited by strange beings, the title could hardly be deemed inappropriate!

For obvious reasons Drake, Berkner, and their colleagues at Green Bank were in no particular hurry to announce their plans. Neither, of course, could their intentions be concealed indefinitely. In September 1959 Cocconi and Morrison, long renowned for their interest and theoretical work on extrasolar planets as potential life centres, published an article in the magazine *Nature* entitled 'Searching for Interstellar Communications'. In this article they suggested that a scheme of experimental work, along lines similar to Drake's, should be put in hand as a matter of urgency. They concluded their article as follows:

The reader may seek to consign these speculations wholly to the domain of science fiction. We submit rather that the foregoing type of argument demonstrates that the presence of interstellar signals is entirely consistent with all we now know and that if signals are present the means of detecting them is now at hand. Few will deny the profound importance, practical and philosophical, which the detection of interstellar communication would have. We therefore feel that a discriminating search for signals deserves a considerable effort. The probability of success is difficult to estimate; but if we never search the chance of success is zero.

In the light of this article Drake and his colleagues decided the moment was opportune. The news was well received, and considerable interest was immediately aroused. There were, of course, dissident voices, and this had not been unforeseen. Now, however, Otto Struve had been appointed as full-time director of the National Radio Astronomy Observatory. Struve lost no time in declaring full support for Drake, but at the same time he did not attempt to minimize the difficulties. He freely conceded how remote were the chances of making a contact in the initial stages of the programme. 'But,' he concluded, 'there is every reason to believe that the Ozma experiment will ultimately yield positive results when the accessible sample of solar-type stars is sufficiently large.' To Struve, Project Ozma was no single desultory experiment by a handful of dedicated men – it was the shape of things to come!

Drake, despite his enthusiasms, kept tight hold on the reins. Communications, he suggested, might be obtained from a quarter of all stars – or from less than one in a million! Background noise and other natural factors over which it was impossible to exercise control would, he predicted, prove the main limitations.

Cocconi and Morrison, who stressed the factors in favour of listening on 21 centimetres, had also envisaged an alien philosophy decreeing the use of alternative frequencies. Drake, however, decided in the first instance to confine his attentions to the 21-centimetre frequency. Other considerations apart, the fact remained that the 85-foot antenna of the Green Bank radio-telescope was well suited to this channel.

The actual announcement of Project Ozma had been preceded by several months of organization and preparation. Although the basic equipment of the observatory was suitable, a scheme of this nature could not be made ready overnight. Additional refinements to the existing radio receivers had first to be made in order to render them suitable for reception of intelligent signals. These receivers, designed expressly for orthodox radio-astronomical research, were already extremely sensitive, and it was estimated that the cost of modification would only be about $2,000.

Economic considerations, however, were by no means unimportant. To this very modest amount of money had to be added the cost of employing additional trained personnel and the acquisition of other items of auxiliary equipment. In addition there had to be considered the general costs of employing the observatory and its great radio-telescope for several months on a programme wholly divorced from the purposes for which it had been created.

In spite of these obstacles Berkner pressed ahead. He was by no means unaware that the very orthodox sections of scientific opinion might severely criticize and castigate him for proposing to spend money and tie up valuable equipment on what to many would seem a hair-brained scheme. All this, of course, was prior to the publication of the paper by Cocconi and Morrison, which as it turned out smoothed the path for Project Ozma.

Berkner felt at this point that the support of an astronomer of considerable repute was essential if he were to carry through with the scheme. With this in mind he sought out Otto Struve to whom he explained the project in full. Although Struve was an orthodox astronomer he had always been favourably disposed to the subject of extraterrestrial life. Indeed, he had shared Drake's interest in this for the greater part of his life, and in his youth Struve had read avidly all that Percival Lowell had written about Mars as a possible abode of civilized beings.

Berkner's negotiations eventually bore fruit for on 3 May 1959 there came the official announcement that Struve had been appointed Director of the National Radio Astronomy Observatory. Berkner could at last relax, knowing that the extremely conservative astronomical circles would be at an obvious disadvantage when the announcement came.

Struve had never sought to conceal his belief in the existence of other solar systems and other civilizations. Early in 1960, with an eye to the advent of Project Ozma, he voiced the belief, based on the retarded rotation rates of several stars, that in the Milky Way alone there were probably no less than *50 billion stars* having planetary systems. Moreover, he added, 'It is probable that a good many of the billions of planets in the Milky Way support intelligent forms of life – I believe that science has reached the point where it is necessary to take into account the action of intelligent beings in addition to the classical laws of physics.'

On 8 April 1960, Drake set the great experiment in motion. Those present that day stood in the presence of history. For the first time in all the long and eventful lifetime of our planet an attempt was being made to contact extraterrestrial living creatures. A new epoch had begun.

The first problem, that of wavelength, had been easy to solve. The second, which was equally important, was slightly less straightforward. Which stars should be selected as the object of the experiment? A list of several could easily be drawn up, each having a good claim. However, with limited resources, limited time and only one radio-telescope available, no more than two stars could possibly be considered initially.

The penultimate selection comprised the stars Epsilon Indi, Epsilon Eridani, Tau Ceti, and Sigma Draconis. It then had to be decided which two stars were to be chosen from this list. The choice did not prove unduly difficult. From the latitude of Green Bank, Epsilon Indi lay too far to the south. Sigma Draconis, lying in the sprawling, northern-circumpolar constellation of the Dragon, was well placed but 18·2 light years remote and therefore less favourable as an initial prospect. Epsilon Eridani and Tau Ceti were both well placed and their distances (approximately 11 light years) were regarded as acceptable.

Drake calculated that intelligent signals emanating from planets of these two solar-type stars could be received at Green Bank by the 85-foot antenna if they were produced by a megawatt transmitter working into a 600-foot aerial system. To be picked up successfully, however, the signals would be required to possess a very narrow bandwidth.

At approximately 4 a.m. on 8 April 1960, our world entered a new age without knowing it. In the control building necessary and final adjustments were made to the ultrasensitive receivers, and current flowed through the complex circuits. Above, under the eternal stars, the great 85-foot dish swung slowly in the direction of the star Tau Ceti, newly risen above the mountains in the south-east. Once 'on target' the clockwork master-control was adjusted so that the great antenna would faithfully follow the star in its passage across the sky.

Time passed. Daylight soon began to flood the eastern horizon. The sun rose, climbing steadily higher into the brassy sky. Hour followed hour. No unusual emissions were received.

The pen of the recorder produced merely the small irregular variations of the 21-centimetre band, the loudspeakers broadcast only the soft background 'breath' of space. In the clear blue sky of the afternoon Tau Ceti was setting even though it was still invisible in daylight's harsh glare.

At last Drake gave the order to direct the antenna at Epsilon Eridani, the programme's other star. The great dish released its 'hold' on Tau Ceti, traversing in an easterly direction until it pointed towards Epsilon Eridani, still invisible in the bright sunlight.

Almost at once it happened! Without warning, before the loudspeakers had even been recoupled, the recording needle began to oscillate violently. So strong indeed were the incoming pulses that the needle actually swung off the paper entirely. After adjustment the needle was seen to record a succession of pulses at the rate of approximately eight per second. Moreover, these were regular and evenly spaced, and there could be very little doubt that they were signals of some sort. The loudspeaker, again operational, verified the signals, which were precise, loud, and clear.

Could it be possible that at the first attempt, only a few hours after the start of this novel experiment, success had been achieved? Were these really signals coming from an unknown planet orbiting the sun known to men as Epsilon Eridani?

Drake was cautious. Premature announcements, which would later have to be rescinded, would make a laughing-stock of himself, Struve, and the entire project. As a first precaution he ordered a thorough check into the working of the equipment to ensure that the apparent signals were not indigenous, originating perhaps within the receiver itself. This possibility was soon eliminated. Whatever their nature these were signals arriving at the great antenna from an *external* source.

The next logical explanation was that the signals were terrestrial, originating from somewhere outside the observatory. Despite the observatory's carefully chosen location and the restrictions that had been placed on electronic equipment in the vicinity, this was a possibility that could not be entirely ruled out.

Nevertheless, the available evidence indicated, or seemed to indicate, that the signals did originate in the relatively minute region of the sky occupied by Epsilon Eridani. There was one simple way to refute or verify this, and that was to swing the antenna away from the star. If the pulses continued then their source must obviously be elsewhere. Unfortunately before this could be done the signals ceased of their own accord, but a fortnight later they began again. At once the huge antenna was steered away from the star. As Drake and his colleagues had feared the signals continued with undiminished strength. They were after all terrestrial in origin!

It would hardly be true to say that the Green Bank team was surprised, although inevitably some disappointment was felt. All had realized that success so early was highly improbable. Nevertheless hopes had been stirred. During the ensuing six months the mysterious signals continued to be received at irregular intervals. Eventually their source was discovered – radar apparatus carried in military aircraft!

Throughout the months of May, June, and July 1960, the great radio-telescope at Green Bank continued to track the two chosen stars. The period was uneventful. There were no further false alarms. Nothing was received that could possibly be interpreted as a meaningful message from a remote star world. There was little genuine disappointment. Project Ozma had always been envisaged as merely the first tentative probing. Other increasingly sophisticated programmes of search could later be initiated, based on the experience already gained. However, a number of matters had to be settled, not the least being the decision to use the 21-centimetre frequency channel. Was this after all the one most likely to bring success? The precise type of receiving equipment necessary would also have to be considered as well as the extent of future star-scanning programmes.

This chapter could not be considered complete without at least a brief survey of the equipment used in Project Ozma. The heart of the Green Bank experiment was, of course, the receiver, but before looking at this in detail it might be advantageous to list the essential requirements of the apparatus employed.

A radio receiver intended to intercept artificial radio emissions from outer space should

1. Cover the frequency range (1,000–10,000 mc/s). This includes the especially desirable 1,420 megacycles per second (21-cm channel).

2. Be extremely stable.

3. Be capable of operating over an extensive range of bandwidths (even as low as or less than 10 c/s).

4. Have a low indigenous noise level.

5. Be capable of reducing *natural* cosmic noise to a low level.

6. Have the ability to discriminate between local (terrestrial) interference and a possible interstellar signal.

7. Possess a very sensitive and efficient antenna system.

Essentially the Project Ozma receiver or radiometer was an extremely stable narrow-band receiver of the superheterodyne type operating in the region of 21 centimetres (1,420 mc/s). The tuning arrangements were such that approximately 100 cycles per second of bandwidth per minute was covered over a total bandwidth of 400,000 cycles per second.

The incoming signal was first amplified and then passed through no less than four frequency changers. The first of these was a quartz-crystal oscillator of very high stability that in conjunction with a frequency multiplier produced a frequency close to that being received. This was then superimposed upon the incoming signal. The output frequency of the quartz-crystal oscillator was 1 megacycle per second, which was raised by the frequency multiplier to 1,390 megacycles per second. Elaborate precautions were taken to ensure the stability of this oscillator, the permitted latitude of which was only one part per billion! In the first instance it was housed within an oven running at a constant temperature of 100° F (37·7° C), which in turn was enclosed within another oven. In this manner the crystal was protected from even slight variations in temperature.

Since the incoming signal was at a frequency of 1,420 megacycles per second, the resultant intermediate frequency (I.F.) became 30 megacycles per second (i.e., 1,420 mc/s — 1,390 mc/s = 30 mc/s). This was amplified and passed on to the second changer, a high-stability, variable-frequency oscillator that, working at a frequency of 26 megacycles per second, resulted in a second I.F. of 4 megacycles per second (30 mc/s — 26 mc/s). This too was amplified before being passed on to the next stage where it was blended with the output from the third mixer (another quartz-crystal oscillator) operating at 3,540 kilocycles per second. The third I.F. was therefore 460 kilocycles per second (4,000 kc/s — 3,540 kc/s). This too was amplified and, after mixing with the 459·8 kilocycles per second output of the fourth frequency changer, was passed on for final amplification at 200 cycles per second.

This large number of frequency conversions was rendered necessary by the extremely narrow bandwidth required. The first oscillator also provided, via a marker-frequency generator, weak impulses at a number of selected fixed frequencies. The purpose of these impulses was to determine the exact frequency at which the receiver was operating. This was of particular importance because it permitted the detection of Doppler Shift. As we saw in Chapter 8 this effect could be brought about by orbital motion of a stellar planet with respect to Earth. The axial rotation of a planet (and hence of a transmitter on its surface) could also be expected to produce the effect.

After the incoming signal had been changed in frequency for the fourth and final time and then amplified, it was passed to a subsequent stage within the receiver. This stage contained two filters known as comparison and signal band filters respectively. These were so adjusted that when broad band noise (i.e., natural cosmic background, etc.) reached them, their outputs were rendered equal. These outputs were then fed into a special part of the receiver called the differentiating circuit. Here the output of one cancelled that of the other. Thus net output of the differentiating circuit was zero.

A *narrow* band signal (e.g., an intelligent message), however, was treated rather differently. Such a signal occupies only *part* of the frequency range of the comparison band filter but *all* of that of the signal band filter. In such instances the output of the latter *exceeds* that of the comparison (broad) band filter resulting in an output from the differencing circuit. In this way the receiver responds only to signals of narrow bandwidth.

Today, only a few years after Project Ozma, Frank Drake stresses the need for a new generation of interstellar communication receivers. 'It is now clear,' he said, 'that the large number of frequencies and stars which must be examined in the search for signals demands, for an efficient search, a receiver of very high information capacity. Thus multichannel receivers or the equivalent of great complexity are called for in future searches.'

Success or failure will not be dictated, so far as equipment is concerned, solely by receiver design. Clearly the most advanced and sophisticated of receivers will prove useless if the aerial system feeding it is inadequate.

The Project Ozma antenna was of the steerable, parabolic (dish) type, 85 feet in diameter, at the focus of which were two prongs or 'horns'. The purpose of the latter was to reduce terrestrial interference as much as possible. (See Figure 18.)

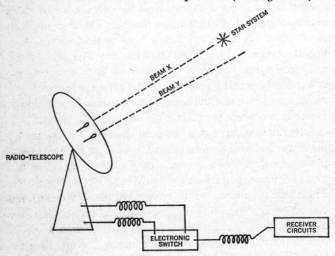

Figure 18

One of the horns surveyed the star under observation (Beam X), the other a portion of the sky lying in close proximity to the star (Beam Y). An electronic switching device connected each prong to the receiver alternately. Thus any radiation picked up from the star would be passed on to the receiver as a series of pulses, the duration of each pulse being dictated by the timing of the electronic switch. At the output end were other devices known as synchronous detectors, which would accept only pulses synchronized with the electronic switch.

Normally interference was picked up by the horns, that is, it was not reflected to them by the dish. Thus it affected both horns to the same extent, and consequently a steady flow of interference passed to the receiver irrespective of the alternations of the electronic switch. However, the synchronous detectors, since they responded only to pulses of specific duration, blocked this steady interference entirely.

It is felt that most of the really large antennas in existence at the present time are too small to be regarded as wholly satisfactory for purposes of interstellar communication. This does not mean that they are *totally* inadequate. However, truly optimum conditions cannot be said to exist if the antenna system is in any way inadequate.

Most antennas over 100 feet in diameter were designed for use at relatively low frequencies and are therefore basically unsuitable for use in the high-frequency range at which interstellar communication is involved. The 600-foot antenna intended for Sugar Grove, West Virginia, might have proved well suited for this purpose. Unfortunately the larger the instrument the greater become the design and constructional problems. This is especially true in the case of the steerable dish type.

In Chapter 8 we gave some ideas of the distance into space that radio-telescopes of various sizes might be expected to penetrate. It might be a good plan at this point to repeat these figures and to give the numbers of separate stars within range of the instruments. These data are supplied in the accompanying table.

Size of Radio-telescope antenna (Feet)	Distance Penetrated (Light Years)	Stars Within Range
85	8.5	6
250	25	60
600	60	1,200
1,000	100	10,000

TABLE 7

Theoretically, therefore, the 85-foot antenna of Green Bank was at a disadvantage, since both stars of Project Ozma were 11 light years remote. The use of a larger dish would certainly have been advantageous.

It is acknowledged that optical telescopes have reached a more highly advanced stage of development than their radio counterparts. This is hardly surprising when we consider that the latter have only very recently made their appearance on the astronomical scene. The final arbiters so far as size is concerned are money and technical know-how, both of which apply to large optical instruments as well.

It has been suggested that orbiting radio-telescopes of vast dimensions might one day be built. In view of our rapidly developing space technology this seems a reasonable premise. Nevertheless, the technique would almost certainly present its own particular brand of difficulties, not all of which might be immediately apparent.

Project Ozma was entirely a 'passive' operation, that is, a watch was kept for intelligible signals but no effort was made, so far as is known, and has yet been countenanced, to *send* messages.

Though transmitters for this purpose may be more expensive to produce and run than receivers, they should on the whole prove more straightforward from a design point of view. With receivers, as we have just seen, it is necessary to provide elaborate devices for filtering out unwanted cosmic background and terrestrial interference. No such auxiliary circuits are required in the case of transmitters. The basic requirement is a fairly normal high-frequency transmitter of considerable power, capable of radiating signals in the 1,000 to 10,000 megacycles per second range. It must be extremely stable, that is, the carrier wave it transmits must be of a very constant frequency and the bandwidth of the transmission should be narrow. It should also have an antenna system capable of being directed exactly at the target star and of sending out a very 'tight' beam in that direction.

We can feel quite certain that space itself presents no obstacle to the passage of radio waves since we are already receiving these in goodly measure from objects of almost incredible remoteness. Admittedly, of course, these are natural emanations, not artificial transmissions.

So the celebrated Project Ozma passed into the limbo of history. After three months the watch on the stars Tau Ceti and Epsilon Eridani was discontinued so that the radio-telescope and its associated equipment could be employed on other research programmes of a less speculative kind. Since the observatory had been constructed primarily to carry out such programmes this was eminently reasonable.

In February 1965, a great new telescope was completed at Green Bank. The diameter of the dish in this case is 140 feet, and at present the instrument is the world's largest equatorially mounted radio-telescope having a solid-surface reflector. Theoretically it should be able to penetrate 14 light years into space, bringing approximately 30 to 35 stars within range. This instrument used in an experiment similar to Project Ozma would, therefore, have a distinct advantage over the earlier 85-foot telescope.

Such, then, was the genesis of interstellar communication. Project Ozma finally removed the concept from the realms of pure theory, stripped it of its science-fiction connotations, and presented it to the world as a rational, feasible thing. There can be little doubt that new and more sophisticated programmes of this nature will eventually be initiated. A true awareness of the cosmic status of interstellar communication is at last descending upon our world!

11 Towards a Cosmic Tongue

BY NOW we have seen something of the technicalities and complexities inherent in interstellar communication. These are difficulties that we believe can be solved so long as the requisite degree of determination is shown and the necessary organization is made available.

This is perhaps the moment to pause and assess the position. In the fullness of time, contact is established between ourselves and an alien race inhabiting one of the star worlds. We receive a signal from representatives of this race. We do not know what it means. All we can say with assurance is that on a planet of a certain star there exists an advanced, intelligent race having a technology at least the equal of our own. A reply dispatched from earth can apparently tell these remote beings only the same thing – that around a certain bright yellow star known to us as the sun (and to them no doubt by some other name) there exists at least one planet on which there lives a technologically advanced race.

Neither of our two races, it would at first appear, could hope to learn more of each other's world and civilization. Such an eventuality would be incredibly frustrating and in the circumstances it seems unlikely that further material and financial support for schemes of interstellar communication would be forthcoming.

Our problem is therefore twofold. There are the technical difficulties, massive and abundant enough by any standard, and there are also the problems inherent in establishing a common means of understanding. Somehow we (and our cosmic friends) must hit upon a means of conveying our thoughts to each other. The ability to transfer mere pulses of electrical energy through space is not in itself enough.

One important point remains in our favour. We are dealing with a race possessing a very high level of intelligence. Thus to some extent the position will be easier than that confronting explorers in parts of Africa and South America two centuries or so ago. Despite actual physical contact, the sheer backwardness of many native tribes made progress towards a common means of expression difficult. Although no physical contact is yet possible with star peoples, an advanced intellectual state must be regarded as a distinct advantage.

A common language must in the beginning be ruled out. It is, however, possible, as we will seek to show, to receive (and impart) considerable information without a single word in common. The answer is basically simple even if the realization is difficult. There is certainly no quick and easy road to success here. All the virtues and qualities so vitally necessary in the technological field will be required in equal abundance. We would remind the reader that these are points we have stressed from the outset. Interstellar communication and interstellar travel are unlikely ever to be spheres of human endeavour suited to the pessimist, the doubter, or indeed to any person of faint heart. It is only for those who truly believe in our race and its future, for those who today believe not just in tomorrow but the day after – and the day after that!

The underlying principle in the medium of communication we are about to discuss is the production of some form of pictorial message conveying a clear, unambiguous meaning. We will assume that alien beings possess the power of sight since this is a sense of such obvious importance. The transmission, therefore, of a form of picture seems an eminently suitable way in which to pass intelligence from one star system to another. It must be made very clear that this would not be a transmitted picture in the television sense. If this were possible the scope would obviously be tremendous. In effect a window would have been created through which we might watch the activities of alien creatures on remote star planets. It may well be that in the far future this will prove to be a feasible proposition. So far as we are concerned, however, such a scheme must be regarded as part of the fruits of a new and as yet undreamed of technology. Have other races in the universe already developed television on the cosmic scale? It is perhaps possible, and therefore presumably just possible that from time to time traces of these attenuated signals reach our planet. If this is true we can only regret our inability to detect and use them.

Our pictorial message will not be something that makes its appearance on a television screen, nor will it in the accepted sense of the word be a picture. Rather, it will be a caricature of beings who live on a remote world and of their civilization.

It is essential that coded signals differ markedly from natural cosmic emanations and from forms of local interference, for if they do not they might easily remain undetected amid the general background babble. A suitable coded signal must be one that will stand out. At the same time the coding must not so modify or alter the basic signal as to render it indistinguishable.

A favourite and by now well-established idea involves the use of a grid or framework for our picture, in conjunction with a system of communication based on prime numbers. The essence of mathematics would, one must suppose, be the same to all intelligent civilizations irrespective of their form or distance from us. Prime numbers should therefore be understood for what they are by intelligent races throughout the galaxy. Thus we have a common factor, which is something so utterly essential to our problem.

Suppose, for example, that from the environs of the star Epsilon Indi we intercept a series of pulses and gaps. Two points gradually become apparent. The gaps are invariably multiples of the pulses, and furthermore the message is transmitted every 20 hours. The first fact clearly denotes a system of coding, whereas the second might well indicate the duration of the planetary day. If we write down an X for each pulse and an O for each gap, then our message from space looks like this:

XOOXOOOXXOXOOXXXOXXOOOOXOXOXOO

Suppose now we find that such a notation is comprised of not just a mere 30 characters but, say, 187 characters. This figure represents the product of the prime numbers 11 and 17. If then we erect a grid having 11 units on its vertical axis and 17 on its horizontal axis, we are in a position to receive an intelligible message. Let us consider a very simple example (see top of page opposite):

oooooooooooooooooooooxoooooooooooooooooxxx

ooooooxooooooooxooooxoxooooooooooooox

ooxxoooooooooooooooxxxooooooooooo

ooox ooooooooooooooooooooooooooo

oooooooox ooooooooooooooooooooooooo

ooooooooooox ooooo

This contains 187 symbols but as it presently stands it is apparently devoid of any meaning. Let us now construct our 11 by 17 unit grid and on this place the symbols of the message. (See Figure 19a.) At first glance it might still appear meaningless.

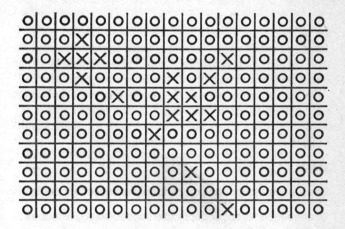

Figure 19a

If, however, on an identical grid we leave blanks for the O's and substitute dots for the X's we get the effect shown in Figure 19b.

Is it really stretching belief too far to suggest that this represents a sun and its four attendant planets? The message can be interpreted as follows: 'Epsilon Indi has a system of four planets.' The message originates on the second planet, which is, therefore, the home of an intelligent race having an advanced technology. We must stress that this represents a very simple message and

187

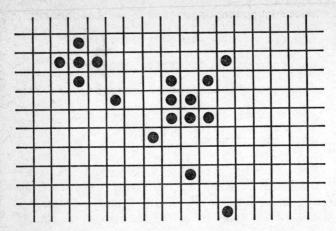

Figure 19b

that of course no attempt has been made to keep to scale. A central sun would normally be very much larger than any of its attendant planets. It is not likely that all planets would be of the same dimensions or have the distances between them identical. We might also query whether the people of an alien civilization would choose an arrow-like symbol to denote their planet. At the moment these things do not matter since our aim is purely to illustrate the principle of the method.

START

OOOOOOOOOOOOOOOOOOOXOOOOOOOOOOOXOXOOOOOOOOOOOXOX

OOOOOOOOOOXOOOOOOOOOOXXOXXOOOOOOOXXOOOXXOOOOOX

OOOOOOOXOOOXOOXOOOXOOXOOOOOOOOOOOOOOOOOOXOOOXOO

OOOOOOOOOOOOOOOOOOOXOXOOOOOOOOOOOOOOOOOOOOOOOOX

OXOOOOOOOOOOOOOOOOOOOOOOOXOXOOOOOOOOOXOOXOOOOOO

OOOOOOOOOOO FINISH

Figure 20a. Portrayal of erect biped as conveyed by a pulse-gap signal of 247 characters. This represents the product of the prime numbers 13 and 19. If 13 is chosen as the vertical axis the result is meaningless. When, however, 13 is made the horizontal axis the result is very different indeed. (Fig. 20b.)

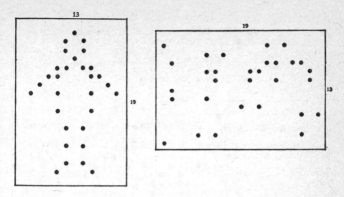

Figure 20b

We would naturally expect a real message to contain a greater number of units than 187 and this necessitates the use of a much larger grid. Frank Drake of the National Radio Astronomy

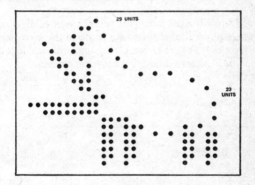

Figure 20c. An advanced intelligent creature on a heavy-gravity planet. In essence the being is a quadruped, squat and low in stature. Since the four normal limbs are necessary for locomotion, an additional pair have evolved that function as arms. Representation of such a creature on a 23 by 29 unit grid.

Observatory has in fact already drawn up a most excellent imaginary signal from space that is comprised of some 1,271 units. Since 1,271 is the product of the prime numbers 31 and 41, a grid having these relative dimensions was used. The message contained an almost fantastic amount of information, some of it of a most detailed kind.

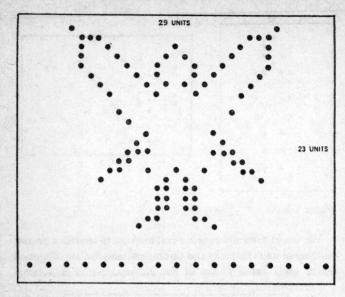

Figure 20d. An advanced intelligent being capable of flight on a planet where physical conditions have rendered this biologically possible. Not only has the creature wings but arms and legs have developed. It is portrayed hovering above the ground. Representation of such a creature on a 23 by 29 unit grid.

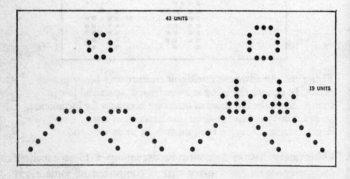

Figure 20e. Simple message from a planet of a variable star. With star at minimum, surface conditions on planet are normal. As stellar activity increases, dormant volcanoes become increasingly active.

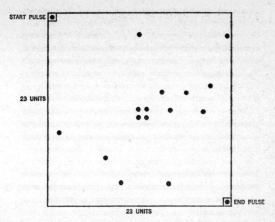

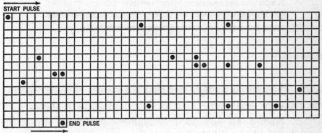

Figure 20f. An imaginary coded pulse/gap message such as might be sent from Earth to reveal to the inhabitants of a planet of our nearest stellar neighbour, Alpha Centauri, the source of this and subsequent communications. The lower illustration portrays a continuous strip of tape 529 units long. When transferred to a 23 by 23 unit grid (upper illustration) we see the constellation Cassiopeia and parts of neighbouring constellations. Heavily accentuated to indicate point of origin, we see the sun as a bright star within Cassiopeia. (It will be of interest to compare this with small star chart, p. 163.)

Drake, having created the 'message', sent copies of it to a number of his colleagues. However, he elected to send not the finished 'picture' but merely a continuous tape of the 1,271 0's and 1's, which is the form in which a genuine message might come from a recording instrument. (See Figure 21.)

Only one of Dr Drake's colleagues succeeded in breaking down the message, which purported to come from a world in which the dominant beings are erect bipeds, mammalian, and of two

```
1000000000000000000000000000000000000000010000111000000000000100
0000000010001000000000100010000000000000000000000000000000000010
0010000100000010000000000010000000000001000100000100010010010000
0100010001000000001110000000000000010000000000001000000000000000
0000000000000000000000000000000000000000000000000001000000000010
0010000011000100000000000000000000000000000000000000000000011000
0110000110000110000110000100000000010010010010010010010010010010
0101010010010000110000110000110000110000110000000000001000000000001
1111010000000000000000001000000000001000001000000000001011011
1001000000000000111110100000000000000000000000000000100000000000
0000001000100111000000000000101000000000000000101001000011001010
1110010100000000000000101001000010000000000100100000000000000000
0100010000010000000000011111000000000000011111000000111010100000
1010100000000000101010000001000000000000010001010000000010100010
0000000000000000010001001001000100110110011101101101010000010001010
00101010100010001000000000000000000010001000100010001000100010000000
1000000000000111000001111100000111000000001111101000000010101000001
01000001000100000010000000001000001000011100001100001000001000001000
0000010000010001000100010000010000010000110000100001000010001000100
0100000100000110000000011000001101100011011000001100111
```

Figure 21

sexes. Their sun is a single star having a system of eight planets, the message originating on the fourth of these planets. The beings possess an advanced technology evidenced by their having visited the third planet in the system, which is water-covered and contains marine life. Symbols denoting hydrogen, carbon, and oxygen are shown, inferring that the chemical basis of life there is akin to that of our own. A form of scale, the units of which are based on the wavelength of the received signal, further indicates that the beings are $7\frac{1}{2}$ feet in height.

This message, therefore, describes a planet in many respects similar to our own. Since its occupants grow to a greater height, a slightly lower gravity and, therefore, a smaller planet could reasonably be inferred. The fact that they are acquainted with the more intimate surface details of a neighbouring planet indicates they have had interplanetary travel for some time and can thus be regarded as ahead of us technologically though not to a tremendous extent. (See Figure 22.)

It has been suggested that a very simple pilot message containing implicit instructions on how to treat and decode a more complex and informative transmission might first be broadcast. This seems a not unreasonable idea in view of the fact that a very

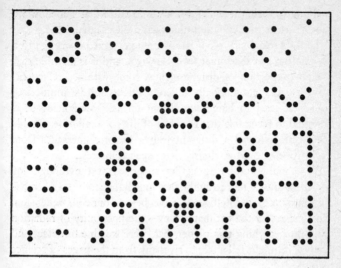

Figure 22

highly encoded message might be difficult to distinguish from background noise.

Pictorial contacts based on prime numbers might even be used to render us familiar with the rudiments of an alien language. The difficulties would obviously be very formidable, the greatest handicap of all being the time factor. A *single* contact with a planet of Alpha Centauri (the nearest star system to the sun) involves a period of 8·6 years, that is, 4·3 years for our signal to reach a Centauran planet and 4·3 years for a reply to reach us. It is this massive, unavoidable time and distance combination that constitutes the great barrier between our own civilization and all others lying out among the stars.

By way of analogy it might be said that at present we stand on a plain. Just ahead lies a range of foothills that we can scale. This is the near future – the already dawning era of interplanetary travel. But beyond, on the distant horizon, lies a great sweeping cordillera whose tumbled, jagged peaks soar into the sky. This is the far future – the incredible era of interstellar voyaging. On the other side of the great mountains may exist another people of whose form, civilization, and tongue we know nothing. We believe they have the means of contacting us, may indeed already be trying to do so. Since we are all too aware of our complete inability to scale these awful peaks we realize that somehow we must learn to intercept and understand their signals.

Drake's imaginary pictorial message portrayed a world and a race not greatly dissimilar from our own, and it is possible that a high proportion of distant planets come within this general category. Planets markedly different might well be inimical to life or at least to higher intelligent forms. Nevertheless it is essential to recognize that the road of life on some other worlds may have followed a different route. Pictorial messages from such worlds would obviously be of special interest.

We should not be altogether surprised at the thought of societies, much more highly developed than our own, transmitting systems of intelligence as yet beyond our comprehension. This makes for the ultimate irony – the possibility of radiation even now reaching this planet that, were we able to detect and analyse it, might tell a tale to rival any from the pages of science fiction.

The idea of an interstellar communication system based on the transmission of vision-modulated signals (i.e., television) may have occurred to the reader, and indeed we have already touched upon the subject in this chapter. The advantages, of course, are obvious. It must, however, be apparent that our own technology is as yet quite unable to cope with such a concept. The difficulties and problems would be vast, but the magnitude of these should not tempt us into the belief that civilizations greatly in advance of our own may not have made considerable headway in this respect. To suggest that some day a faint, fleeting scene set on the surface of an alien world several light years remote will for a moment or two flicker uncertainly on a cathode-ray tube here on earth seems to be something straight from the realms of sheer fantasy – and yet who would be so bold as to state that this could never be!

We must not, however, allow our thoughts to flow too freely along these lines. A television link between worlds owing allegiance to different suns seems almost too perfect and too sophisticated. All we can do is to accept the *possibility*, however remote and fantastic it may now seem.

A related possibility is that some of our 'near' cosmic neighbours could by means of interstellar probes receive and interpret something of our own domestic television transmissions. This aspect, the injection of alien probes into the solar system, is dealt with a little more fully in a subsequent chapter.

Because mathematics must represent a common ground among advanced societies, the possibility of imparting information by mathematical means has been favoured for a long time. The application of prime numbers is a case in point and could, as we have seen, be the means of imparting a considerable amount of information. Recourse to pure mathematics would probably be adopted by an intelligent alien civilization in the first instance to indicate to other societies that it had achieved a high intellectual standard. Towards the close of the nineteenth century, when it was thought that Martians did exist, a suggestion was made that on one of the large desert tracts on Earth there should be constructed a vast diagram illustrating Pythagoras' celebrated geometrical proposition. Obviously, visual means of conveying such intelligence would be of no use whatsoever in communications between planets of different stars.

To convey the basic arithmetical processes of addition, subtraction, multiplication, and division by regular fluctuations in the strength of radio signals would be fairly straightforward. Let us envisage a pen recorder receiving the following signal from a radio-telescope.

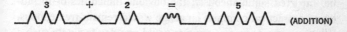

Each unit is represented by a single pulse while 'plus' ($+$) and 'equals' ($=$) can be indicated by a particular short signal showing variations in amplitude. In the circumstances it must be obvious to any thinking beings that the 'radioglyphics' concerned stand for 'plus' and 'equals' respectively. In similar fashion subtraction, multiplication, and division can be portrayed.

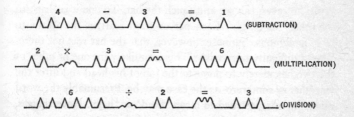

Half ($-\frac{1}{2}$) could be represented by a half pulse.

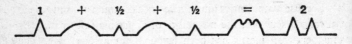

This, however, might easily lead to confusion, and such a system would certainly not lend itself readily to the expression of other reciprocals. There would, therefore, be a clear need to establish a symbol that, placed immediately prior to a number, would indicate that number's reciprocal, for example:

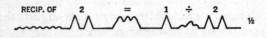

and similarly,

If we take the title of this chapter literally the problem becomes infinitely more complex. As has already been pointed out, the time lag to which interstellar signals are subject is a severe and unavoidable handicap. Even though electronic messages could be imparted simultaneously, the obvious difficulties of formulating a practical common language would remain tremendous. Whether or not supercomputers could one day facilitate such a project is something that only the future will reveal.

If the reader cares to think about the language issue a little more closely he or she will realize to the full the inherent and tremendous difficulties that beset it. Let us for a moment or two consider the case of an English-speaking person and a Spanish-speaking person who are both looking at a hat. Neither person, we will assume, understands a single word of the other's language. Both, however, can see and touch the hat. To one it is 'hat', to the other 'sombrero'. Each person soon realizes the words to by synonymous. Suppose, however, that the hat was not there. Would difficulty then arise? This is unlikely for one or other of the two has merely to point to the top of his head and utter the word hat or sombrero, as the case may be. Presumably the word might be misconstrued to mean 'head' or 'hair'. Nevertheless, were the necessary materials available, a hat could be drawn or even scratched on the ground with the aid of a stick or a stone. If, however, our English-speaking friend and our Spanish-speaking friend were some miles apart and linked only by telephone, difficulties would multiply at once. The English-speaking person could shout the word 'hat' into the microphone indefinitely but he would certainly fail to convey any meaning to his Spanish-speaking friend. There would simply be nothing to link the word hat with the object known in Spanish as sombrero. Remove microphone and earphone from each end of the link and replace these with Morse keys and buzzers. Now the problem is rendered still more acute for the communication link has become merely a telegraph system over which voices cannot be heard. We have in fact reached a state of affairs analogous to that of an interstellar contact. Two parties knowing precisely nothing of each other's language receive only pulses of electrical energy. They are unable to see one another, and so gestures that might convey meaning are out of the question. Here of course the comparison ends. Our terrestrial friends can always learn Morse and so at least receive the actual words. They can then proceed to look these up in an English–Spanish dictionary. Unfortunately there is as yet no celestial Morse Code, no terrestrial–cosmic dictionary!

A communication system between Earth and an inhabited star world, combining prime-number 'visual' transmissions *and* a voice-modulated sound broadcast, could perhaps in time achieve a measure of success. We could, for instance, transmit on a prime-number grid the representation of one specific object. At the same time the voice channel could repeat over and over again the word for this object in one of the terrestrial languages. In time we might hope that the aliens would re-transmit the 'picture' and accompany this with continual repetition on sound of *their* word for the object. This would undoubtedly be a slow and ponderous undertaking even assuming voice-modulated transmissions could travel such a distance and remain intelligible.

We have then some idea of the truly immense obstacles in the way of building up a common language. We have thought in these instances only of a single word. But language, any language, is not just a single word – it is thousands of words. Also, these words do not all signify specific material objects. All words are not convenient nouns. We must consider the various parts of speech, all the verbs and adjectives, all the pronouns and adverbs. We must think of grammar, idiom, and expression in general.

The problem of language in the interstellar dimension is really one of two parts. Words must first be conveyed from one star system to another. This having been done (no mean achievement!), they must then be translated.

A *written* message we must suppose could be transmitted between planets of different stars using the prime-number method previously outlined. This would presumably be more involved than picture transmission, and its success might largely depend on the nature of the symbols or hieroglyphics in the language concerned. If this could be achieved by an alien race we would then have a few words or a simple sentence made up of characters totally unknown to us – an impasse, it would seem. Though no one would for a moment underestimate the difficulties, this impasse need not necessarily exist. Already on one or two occasions inscriptions in languages, all previous record of which had been lost, have been successfully translated.

An interesting semi-parallel is provided by the case of the famous Rosetta Stone. This is a block of basalt stele found in 1799 by a French officer named Boussard. It was discovered near Fort St Julien, which lies four miles to the north of Rosetta, a town located at the western mouth of the Nile River. The Stone, which is inscribed in hieroglyphics, demotic, and Greek, now reposes in the British Museum. The existence of demotic and Greek inscriptions enabled Champollion to decipher the hieroglyphics. Armed with this key he was then able on an extensive scale to set about translating the hieroglyphics of ancient Egypt.

The reader will at once point out, and quite correctly, that unfortunately no celestial Rosetta Stone has been found wandering about in space to provide the convenient key to an alien language. However, there are mathematicians who believe that a certain type of symbolism could be developed that might in a sense provide something of a substitute.

One concept is to employ unmodulated radio signals the wavelength and duration of which vary. These it is thought might be combined in such a way as to constitute definite words or ideas and in this manner vocabulary and grammar could perhaps gradually evolve. The whole business is extremely involved and would seem to depend to a large extent on aliens employing mental processes akin to our own.

In general, however, the concept of a tongue common both to ourselves and to an alien race whose form, civilization, and basic philosophy might differ so radically from ours is hard to accept. To beings our intellectual superiors by several millennia the problem may seem considerably less acute. On such a race would inevitably rest the onus of trying to make the occupants of primitive Earth comprehend. From a distance of several light years even this might prove too great an obstacle for even the most enlightened of cosmic civilizations. This is not intended as a disparaging comment on our society on Earth; we must remember that the age of a civilization, like the age of an individual, is a factor over which there is no control. Our terrestrial society has on the cosmic time scale moved a certain distance. Had our world been born sooner, had life started on Earth earlier, we would today have found ourselves further along that scale and consequently in a more enlightened state. The fact that we are not is hardly due to shortcomings on our part.

It seems therefore that the transmission and reception of prime-number pictures (or some variation on this theme) affords the best hope. If by this method aliens could in time instruct us in the rudiments of interstellar television and the reception of sound-modulated cosmic broadcasts a reasonable foundation might be created towards learning their language. This would certainly be superior to the use of mathematics or of logic. In this field all things are very relative, and what is easier than something else will still be infinitely difficult to accomplish!

12 Other Techniques

THE title of this book might seem to imply that forms of communication between our world and other solar systems *must* be of an electronic kind. This is not necessarily so, although other forms of contact are probably less likely. In fairness, therefore, one chapter at least should be devoted to some other possibilities. Indeed this is essential if an overall picture of interstellar communications is to be drawn.

The other most probable method would seem to involve the use of laser beams. Lasers represent a very recently acquired facet of technology, and incredible progress has already been made in the field. Generally we see interstellar communication as an exercise in the use of a particular portion of the electromagnetic spectrum. The use of lasers parallels this almost exactly except that the portion of the electromagnetic spectrum involved in this instance lies in a much higher frequency range – that of visible light.

It may be that the possibilities of using light beams as a medium will already have occurred to the reader. At a first glance it might seem that light would have much to commend it, since after all man first became aware of stars and distant galaxies by virtue of the light waves these objects emit. Only in comparatively recent times have we become aware that many of these bodies emit radio waves as well. Unfortunately normal light propagated by artificial means would be of little use as a medium of communication over interstellar distances. Indeed it is doubtful whether it could even be used satisfactorily between our own planet and Mars, although between Earth and the moon its use might be a feasible proposition. In the latter instance, however, this could only be as a novelty since radio communication would possess a clearer advantage.

Let us consider the subject of ordinary light for a moment or two. Those of us in Europe able to recall the early years of the Second World War will no doubt remember the use of searchlights to detect hostile aircraft. These produced immensely powerful beams that could easily illuminate aircraft at a height of several thousand feet. The light was apparently confined to a well-defined beam – in other words, it did not noticeably spread out. Only in a relative sense was this true, and in fact the farther the beam travelled the greater became its spread. As a consequence its effective range was considerably restricted. (See Figure 23.)

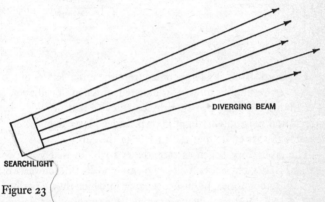

Figure 23

It will be observed that rays of light coming from the precise centre of the searchlight are at right angles to it. From all other points the angle is in excess of 90 degrees, and the nearer these are to the perimeter of the searchlight face the greater is the angle. In other words the rays are spreading out and thus the beam, which is a combination of all the rays, is also spreading out. The effect on the initial portion of the beam is so slight that there would seem to be little divergence of the rays. The farther they travel, however, the more apparent does the effect become.

If the individual rays comprising the beam could all be rendered parallel to the rays from the centre, a tight, sharply defined beam would result. This in a sense is what a laser beam achieves.

Ordinary white light is comprised of light at many different wavelengths, and these waves travel in every direction. For this reason it is generally referred to as *incoherent light*. The light produced by a laser is for the most part of one wavelength only. The waves are, therefore, unidirectional and have the effect of reinforcing one another. As a consequence the resultant beam is straight and very narrow over long distances. This is an example of coherent light. (See Figure 24.)

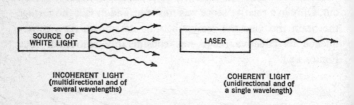

Figure 24

The word 'laser' is a term coined from the initial letters of the words '*l*ight *a*mplification by *s*timulated *e*mission of *r*adiation', a brief and strangely prosaic title for a device that provides the purest and most intense light known. Indeed laser light is comparable to that emitted at the surface of the sun.

The laser story began as recently as 1958. In that year two United States physicists, C. H. Townes and R. N. Schwartz of the Bell Laboratories, published a paper in which they described how a laser beam might be created. It was, however, left to another American, Dr Theodore Maiman of the Hughes Aircraft Company, to give the idea practical expression.

The underlying principle of the laser is comparatively simple. At the centre of the device is a thin rod of synthetic ruby crystal, the ends of which are so treated that one acts as a total reflecting mirror and the other as a partially transparent surface. The rod is surrounded by a form of flashtube not unlike the type used in high-speed photography. When the flashtube is discharged a very intense light is generated, the effect of which is to excite the electrons within the crystal rod. Ruby is a form of aluminium oxide in which a few of the atoms of aluminium are replaced by those of chromium. The chromium atoms, unlike those of aluminium, possess electrons that are not 'locked' within the crystal lattice and can as a result be raised to a higher energy state, a function achieved by the discharge of the flashtube. As the excited electrons revert to their natural level they emit photons of light. These photons traverse the length of the rod, rebounding between the mirrors. In so doing they cause other electrons that have been raised to higher energy levels to emit light also. The cumulative effect is a torrent of red light oscillating several million times between the mirrors within a few thousandths of a second. Finally the light becomes so intense that it pierces the partially transparent mirror and emerges as a laser pulse of coherent light.

Unfortunately this type of laser device is not capable of continuous operation, that is, it produces *pulses* of laser light but not a steady beam. In 1961, however, a laser producing such a beam was perfected. This differed from the earlier type in that the ruby had been replaced by a helium-neon gas mixture. Indeed a variety of different substances can be used in this role, for example, gallium arsenide (especially useful if the laser is to function at very low temperatures), and carbon dioxide (for the emission of very hot infrared light).

A fascinating experiment involving the use of a ruby laser was carried out in May 1962 when scientists at the Massachusetts Institute of Technology used it to illuminate a 20-mile-wide portion of the moon's surface, a target a quarter of a million miles distant! The resulting reflection from the lunar surface was positively detected by ultrasensitive measuring equipment on Earth. Had a sufficiently high-powered searchlight been employed instead, its beam, due to spread, would have been equal to several times the diameter of the moon (2,160 miles). Thus we begin to see the potential of the laser as a 'light' communicator over cosmic distances.

Already two methods of using lasers in this respect have been suggested. The first envisages the passing of a laser beam through the optical system of a giant telescope, such as the 200-inch reflector at Mt Palomar. The resultant beam would then be 200 inches wide instead of the few inches of a typical laser. The other involves a cluster of perhaps 20 to 30 lasers working in unison to produce a beam 4 inches wide. Unfortunately in each instance the effect of the atmosphere would be to scatter the beam. This particular difficulty could of course be obviated by placing the source of the beam *above* the atmosphere on a space platform or the like. Though such an expedient might be feasible in the case of a laser cluster, its adoption for a 200-inch telescope combination seems at best unlikely! In this instance the use of a much smaller telescope would be called for, though eventually a giant telescope on the airless moon might prove a workable proposition.

Were it possible to use a laser–giant telescope combination, the beam, it is reckoned, would be visible by the naked eye at a distance of 0·1 of a light year. Slight optical aid might extend this to 0·4 of a light year. A laser cluster arrangement, on the other hand, would give an appreciably poorer result – probably something of the order of 0·01 of a light year in the case of the unaided eye and 0·04 of a light year with slight optical aid. The use of a 200-inch Palomar-type telescope at the receiving end would obviously improve matters greatly in the case of both systems, and it is believed that a beam from a laser–telescope arrangement would then be visible from as far away as 10 light years and possibly from 2 to 3 light years in the case of a laser cluster. The interstellar potentialities of lasers are therefore considerable, and though their range, so far as we are concerned, is still barely good enough, our advancing technology should eventually bring about the necessary degree of development. What, then, can we say of civilizations whose capabilities in laser science may be very far in advance of our own? Already laser beams may be probing the eternal dark night of interstellar space.

Another difficulty now arises. Suppose, for instance, that a race on a planet orbiting the star Sirius were to direct a laser beam towards us. Such a planet must of necessity lie reasonably close to the star to be habitable and therefore laser light coming from it would be swamped by the relative flood of light emanating from Sirius itself. If, however, the laser beam were to produce light of a wavelength capable of being *absorbed* by the atmosphere of the star, the spectrum of Sirius would then show a black absorption line where physically none should exist. Before proceeding too far on this theme, however, some explanation should be given for those readers unfamiliar with this aspect of physics.

Most people are familiar with that lovely arc of spectrum colours known as a rainbow. Raindrops in this case are breaking up normal white light into its various components. A similar effect can be produced artificially in the laboratory by the use of a glass prism and a source of white light. (See Figure 25.)

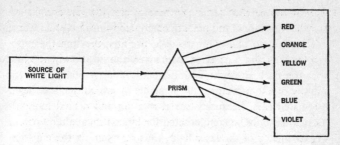

Figure 25

A small pocket spectroscope (of the type used in qualitative chemical analysis) will also produce this attractive band of continuous merging colour. If this is directed at the sun, however, a series of straight vertical lines appears at various points along the coloured spectrum band. (See Figure 26.) These are known as

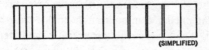

Figure 26

absorption lines and are due to elements within the sun's atmosphere selectively absorbing some of the sun's light. In this way it is possible to determine what elements are present in the sun. Indeed this is the very basis of stellar spectroscopy by which we can determine the constitution not only of the sun but also of stars and other remote celestial objects. Each element absorbs light at a specific point in the spectrum, and thus each line is characteristic of a particular element. Each such line has also a certain breadth. If, for example, a laser generates a beam the wavelength of which would produce a bright (or emission) line in the centre of a black absorption band, an artificial characteristic would obviously have been given to a spectrum pattern. In this way a laser beam could indicate its presence even though it were lost in the flood of light from a nearby star.

This may seem a relatively easy way in which to surmount the problem. Unfortunately there are still difficulties. So far as we are concerned this emission line within an absorption band might, were it too narrow, remain imperceptible. It is an established fact that brightness and breadth of emission lines have an inverse relationship to one another – the brighter the line the more narrow it is and vice versa. Moreover, the position of the line within a dark absorption band would almost certainly tend to wander due to movement of the planet on which the laser beam originated (i.e., because of axial rotation and orbital motion). This could only be compensated for by continual alterations in the frequency of the laser light, thus maintaining the emission line in a constant position.

The potential of the laser so far as interstellar communication is concerned is thought to be considerably less than that of the more orthodox radio. On the other hand many of the possible disadvantages have been suggested by men who are strong protagonists of the latter. However, the problems involved are real enough and should no more be underestimated than those peculiar to radio techniques.

Certainly the position so far as Earth is concerned is that radio methods of interstellar communication hold a clear and commanding lead. The superiority possessed by the latter is particularly apparent in respect of range. Although radio messages could be sent out as far as 1,000 light years, it is doubtful whether their laser counterparts could exceed 10 light years – a hundredfold difference. Background galactic interference is less when the frequency is lower, and this also favours radio. Visible light is, of course, electromagnetic radiation of a much higher frequency. The construction of sufficiently large optical mirrors as receivers also poses an acute problem, whereas extremely large radio mirrors (i.e., dish reflectors) are already in existence. Above all there remains the problem of using a laser beam (producing an emission line in an absorption band) as a means of conveying intelligence.

In enumerating difficulties we must, as with radio techniques, always remember that some civilizations may long since have resolved these. Regrettably, of course, if the people of Earth are unable to overcome receiving problems, no amount of technical sophistication on the part of those transmitting signals, be these optical or electronic, will avail.

It is considered possible that with certain civilizations the advent of lasers may conceivably have *preceded* that of radio. On

Earth radio was the first to appear. Because of this we are much more familiar with it. Lasers are still largely a closed book. We tend, therefore, to regard radio as a much simpler concept, an idea conducive to the belief that in every civilization it must have preceded the laser. This is a doubtful premise and the fact that radio came first with us may well have been a coincidence.

A number of other suggestions have also been made from time to time regarding the way in which alien societies might announce their presence. Some are extremely ingenious but most are highly improbable.

When discussing lasers we saw how the various elements in the atmosphere of the sun absorb light of a specific frequency and thereby give rise to dark vertical absorption lines. If an element not already present in the sun's atmosphere could be injected into that atmosphere, then an absorption line would almost certainly make its appearance in the visible spectrum where none normally exists. The problem here is whether such an element is available. One does in fact exist though unfortunately it does not occur naturally on Earth. This is technetium, which has been prepared artificially in very small amounts. Its presence has not been revealed in the spectra of other solar-type stars though it is thought to exist in stars of spectral group S. For the purposes of the project relatively little of the element would be required – probably about two or three hundred tons. In the case of a rare material like technetium, however, such a figure represents a *vast* amount, the preparation of which would tax the resources of the most highly developed technology. Its injection into a stellar atmosphere would also pose considerable problems. To ensure the appearance of its characteristic absorption line would necessitate the element's being distributed uniformly throughout the stellar atmosphere. In addition its presence would be required to be maintained against the steady pressure of our flowing gases from the star.

Such a scheme could hardly be adapted to transmit information and must be looked upon merely as a sort of marker beacon whereby another civilization endeavours to announce its existence to the universe at large.

The possibility of using a star itself in this context has also been proposed. Fundamentally the idea is straightforward enough, although putting it into effect would certainly raise vast problems. Essentially the concept envisages the placing of a dense cloud of particles in orbit round the star. By so doing the light of that star is partially or wholly obscured at set intervals with respect to points in the plane of star and cloud. Such a cloud would be required to possess a very considerable mass (100,000 billion tons has been suggested!) and since it would be composed of individual solid particles it would be easily prone to disruption.

Probably the most fantastic and certainly the most spectacular idea of all has been suggested by Freeman J. Dyson. This involves nothing less than the total dismemberment of a planet or planets in a planetary system and the utilizing of the material to create a shell completely surrounding the parent star. This shell or envelope, about 10 feet thick, would lie approximately 180 million miles from its stellar 'core'. The exterior of the shell would in these circumstances radiate to the same extent as the star itself – with one fundamental difference: the radiation would be confined entirely to the infrared region of the spectrum. Such radiation could easily penetrate the atmosphere of a planet like our own. Because of this it has been suggested by Dyson that a search be instituted for strong sources of radiation in the appropriate region of the infrared band.

It is rather difficult to take such an idea seriously. Even the most enlightened, sophisticated, and technologically advanced society would find the project of such stark magnitude as to make realization impossible. Nevertheless the Soviet astronomer Shkovsky has shown a willingness to endorse the general principles inherent in the idea, and he has indeed even gone so far as to suggest that since the Great Andromeda Galaxy (the nearest universe to our own) can be surveyed in its entirety, it might be a good idea to look there since a marker beacon so prominent could be detected even at a distance of $2\frac{1}{4}$ million light years!

At this point it is convenient to enlarge upon the possible advantages of using infrared wavelengths. Such a choice may have a more logical basis than has hitherto been supposed.

A rather interesting signalling system has been suggested by A. T. Lawton. This involves the use of super mirrors, each 1,000 inches in diameter, for operation at infrared wavelengths. One of these has a one megawatt laser at its focus, the other a photo-conductive detector with extremely high sensitivity (10^{-15}

watts). The theoretical distance over which such a system would operate is 250 light years. If a 3,600-inch mirror were used the effective distance would increase to over 3,000 light years.

Such a system assumes conditions of near perfection including a noiseless background. This, of course, can hardly be expected in practice. It is believed, however, that in directions of low noise a range of up to 500 light years should be possible. In view of the inevitable time-lag this is not of great concern. The practicalities of the situation demand a much more modest span – something of the order of 10–100 light years.

Should it ever be necessary to communicate over greater distances this might be achieved by the use of more powerful lasers. A 1,000 megawatt laser battery and a 3,600-inch mirror would give a theoretical range of 15,000 light years. This could also serve as a high-intensity *short*-range system for areas of *high* galactic noise.

Earlier we mentioned the logic of using a frequency of 1420 mc/s. The logic behind the use of an infrared system is rather less obvious and could perhaps be attributed as much to economics as anything else. The cost of the type of installation just described would, it is reckoned, be much less than one of corresponding 'radio' power. In the case of a one megawatt laser installation, this would probably amount to less than 5 per cent of its 'radio' counterpart. We may reasonably assume that the keepers of alien treasuries will regard this state of affairs as equally attractive!

It would, however, be wasteful to embark on such a programme in a haphazard way. In other words it is rather pointless signalling a likely star or even 'listening' to it. The initial stage of the programme would therefore involve a survey of likely stars in the immediate environs of the solar system to ascertain which had planets. An infrared laser beam directed towards the sun from a planet of one of these stars would easily be detected by ultra-sensitive equipment. The effect of this would be to indicate an infrared double star where (visually) only a single star exists. For this stage of the programme modest installations would suffice.

Once positive results had been obtained the second stage of the programme could get under way. This would entail the construction of a much larger infrared installation with a mirror in the 1,000–3,600-inch range to continue the survey.

The third stage would follow detection of a signal. The latter would require to be checked several times and decoded (if possible). The search could then be continued to locate other sources.

It is here that Lawton raises a most interesting point – whether or not a decision to reply should be made, since doing so will announce to the galaxy that another highly technological civilization has emerged. The personal opinion of the present writer is that if the situation arose a reply should be made. Risks there may be but little of worth will ever be achieved without some element of risk. As it is, what guarantee have we that our normal high frequency indigenous radio signals have not already been intercepted and monitored by alien civilizations at several points within the galaxy?

Even if an advanced civilization has not elected to transmit cosmic signals of any kind, its presence might still be detected by the 'simple' expedient of eavesdropping, that is, listening to its indigenous radio traffic. This does not necessarily mean picking up its domestic and commercial broadcasts, although a society sufficiently advanced in electronics might presumably manage to achieve this. We might, however, consider radio signals passing between individual planets of a system, communications between interplanetary space vessels, long-range radar, and so on.

It is on the whole unlikely that we could detect in this way the presence of other civilizations even a few light years distant at the present time. Listening must still be carried out from the surface of our planet and the effect of its atmosphere would almost certainly be to nullify such weak signals. Receiving stations on the moon or on orbiting space stations could, however, alter the situation in our favour. Even allowing for this, immense difficulties would remain. Such signals would be extremely faint and therefore easily missed. Indeed they could be detected without their true nature being realized.

Frank Drake, the originator of Project Ozma, firmly believes that a programme of listening along these lines might not be without merit. His suggestion is an extremely good one, almost brilliant in its simplicity. Since signals picked up in this way could so easily be missed or misinterpreted, Drake proposes that an extremely sensitive radio receiver be used in conjunction with a radio-telescope, the latter being directed towards a suitable star (à la Ozma!). The procedure in this instance would not involve listening continuously on 21-centimetres or indeed on any other single frequency. Instead the entire radio portion of the spectrum would be examined, probably in sections of 100 cycles per time. The received noise would be recorded. After this had been done the process would be repeated. If subsequently the two tapes were played back in unison, artificial signals picked up on both would be superimposed and therefore strengthened. With random static this would not be the case. In this way artificial signals could be made to stand out much more clearly.

Picking up indigenous signals from other solar systems can be given a more interesting slant if we also consider the possibility of intercepting signals passing between interstellar space vessels or between such vessels and their home planets. Our detection of alien radio traffic too strong to emanate from points as remote as the nearest stars would indicate that aliens, although perhaps not about to descend upon us, were nevertheless close in the stellar sense. The gradual increase in strength of such a signal coupled with an element of Doppler Shift might be taken to indicate the approach of such a craft.

The use of thermonuclear blasts to indicate the presence of a civilization has also been suggested. On the whole this remains unconvincing. Although the detonation of such a device is by terrestrial standards an awe-inspiring manifestation of physical forces, it is incredibly minute in comparison to the hydrogen-to-helium transition going on in a star. Such artificial blasts even on the outermost planet of a system would to other planetary systems be lost in the glare of the devastated planet's sun. Admittedly, thermonuclear-tipped projectiles could be sent off into interstellar space and then detonated, but it is hard to see what this would achieve. Even if the brief transient flash were spotted and correctly interpreted the only information it could possibly impart would be that somewhere perhaps 5 to 10 light years distant there existed a race with both thermonuclear and potential interstellar capabilities. The exact location of that race would remain uncertain. It is difficult to believe that a race so advanced would resort to such a crude and singularly ineffectual practice.

It would be unfair to close this chapter without a reference to telepathic possibilities. Telepathy is a subject about which very little is known. Indeed no one can really say whether it has a true basis, and many dismiss it as bogus. Even if its validity could be proven this does not mean it could be practised over interplanetary or interstellar distances.

Many examples of supposed telepathic contacts on our world could be quoted. Most of these are at best unreliable and should be treated with the greatest of reserve. Regrettably this is a field in which charlatans have in the past flourished.

During the 1963 International Astronautical Congress held in Paris it was revealed that several Soviet research centres were investigating the physiological basis of thought transference. At this point the words of E. A. Asratjan, a member of the Soviet Academy of Sciences, are worth noting. He said:

> The problem of thought transference over distance is a very complex and controversial one. We will need many more experiments, following the strictest proceedings of experimental design, before we can be fully assured of the existence of this phenomenon. On the other hand we are in possession of certain data which prevents us from denying its existence.

One of the leading Russian research workers in this field is L. L. Vasiliev who occupies the Chair of Physiology at the University of Leningrad. His view is that thoughts may be transmitted over a distance either by a form of energy already known to exist, by one not yet known, or by some other medium unrelated to energy.

Vasiliev carried out a number of interesting experiments involving three persons. One was a subject who believed he was merely undergoing a routine medical check, and the others were a hypnotist and the experimenter himself. The role of the subject was to squeeze a rubber bulb with a regular rhythm, the purpose being to indicate the onset and termination of hypnotic sleep. The hypnotist was located in another room, which did not adjoin the room of the subject. When it appeared that the suggestions of the hypnotist were reaching the subject, the latter was enclosed in a lead chamber hermetically sealed by mercury. The purpose of this chamber was to rule out the possibility of electromagnetic radiation reaching the subject. Despite the use of this chamber, results were still apparently positive, indicating that the thoughts of the hypnotist were not being conveyed to the subject by a normal electromagnetic medium, although the possibilities of the latter being affected by gamma rays, X rays, or waves of considerable length could not be ruled out.

Any theory of long-distance telepathy based on the transmission and reception of some kind of electromagnetic radiation is frequently disputed on the grounds that intensity of electromagnetic waves decreases with distance – or more accurately as the square of the distance. The suggestion has been made, however, that decrease in radiation intensity is not in itself vital so long as it remains possible to distinguish the signal from interference. The relevance of this so far as telepathy is concerned lies in the belief that interference from other mental signals would be low and that the quantity of energy required to actuate the receiving mechanism of the brain would probably also be low. The analogy of the eyes' response to the faintest light has been quoted in respect to this.

Even if we accept the existence of a *prima facie* case for telepathy, the parameter of distance that cosmic contact introduces would not, it seems, be one that could lightly be ignored; this is not distance in the normal sense – this is distance that defeats imagination, distance that even light takes years to traverse.

There is, moreover, a more important aspect. So far our thoughts have dwelled exclusively on the idea of contacts between human beings. What might the position be telepathically between ourselves and alien beings, even if the latter bore a marked resemblance to ourselves? This would be even more valid were the aliens totally different in form and mental outlook. Minds of the same kind might be in tune though light years apart. Minds totally different in bodies equally different might in the metaphorical as well as the literal sense be on entirely different wavelengths.

So far as interstellar communication is concerned telepathy may remain a purely academic consideration – at least, that is, until we know a very great deal more about it. It may be that deep within our subconscious lie thoughts propagated and implanted by other brains. Perhaps the 'tuned circuits' of our brain cells are simply unresponsive to these.

Occasions are known when persons visiting a part of the country or world for the first time are suddenly and acutely conscious that the scene confronting them at a particular moment is strangely familiar. Indeed in some instances they may even be vaguely aware of what they will see round the next bend in the road. Are these thoughts due to the subconscious pick-up of the thought waves of others? Fantastic though this may seem there is an element of credibility about the idea. In some such instances, of course, what may actually have happened is that the scene has become confused with a very similar one seen much earlier in life.

Instances are known of dreams that consist largely of the persistent reappearance of a particular scene. Sometimes the scenes are said to have a distinctly alien appearance, apparently corresponding to no known setting on this planet. It would be all too easy to let our thoughts embrace the intriguing possibilities inherent in this. Dreams, however, are frequently irrational, and this irrationality may extend to the type of scene that appears. Such a scene might conceivably make so great an impression on the mind of the dreamer as to cause its repetition on a number of occasions.

We do not know whether it is possible for an alien race with a highly sophisticated technology to transmit intelligence artificially in such a way that it is picked up automatically by other minds on other solar systems. This seems so much a science-fiction concept that one hesitates even to accept it as a remote possibility. In these pages, therefore, it is mentioned only as a passing thought!

13 From What Far Star?

So far we have restricted ourselves almost entirely to the concept of signals passing between one star system and another. In Chapter 6, however, we looked fairly closely at the tremendous difficulties implicit in interstellar travel and came to the conclusion, rather reluctantly, that for the present and foreseeable future such an adventure for the human race could not be contemplated. It may, therefore, seem odd to consider interstellar probes as a means of expediting interstellar communication, and we must concede that such probes still involve us in most of the complexities and difficulties inherent in interstellar travel.

It is felt by some researchers in the field of galactic communication that the idea of maintaining a host of transmitters over many years in the hope of securing one contact is not only wasteful but downright impractical. Instead the suggestion is made that we do what is already being done with respect to the planets of the solar system, that is, dispatch instrumented probes. These, however, would not merely be probes – objects designed to send back information regarding a planet's topography and atmosphere – but things more in the nature of travelling relay stations.

There is, of course, one very big snag. While it is possible for us to send probes to Mars and Venus and even to the outer solar planets, it is impossible to dispatch such objects to the environs of even the nearest stars. The last sentence is not strictly true. Our contemporary technology *does* permit us to build and dispatch probes that *could* reach such remote worlds. The time, however, that they would take renders the whole concept impracticable. Once again we are confronted with the awesome time and distance barrier that is so integral a part of the interstellar scene.

Let us suppose that a probe were dispatched from our solar system to the nearest star, Alpha Centauri, which is 4·3 light years distant. If the velocity given to this object were 100,000 miles per hour, then well over 285,000 years would elapse before it reached its destination. (See Figure 27a.) Now it seems unlikely that the terrestrial civilization of that era would be remotely interested in a probe dispatched in 1976 – assuming the thing were still functioning. By then interstellar travel should have been developed, and by a strange irony distant descendants of the probe's progenitors might be on hand to witness its belated arrival within the bounds of the Alpha Centauri system. It would certainly be eligible for exhibition in the nearest space museum!

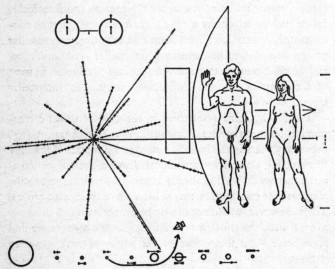

Figure 27a. Pictorial plaque on the *Pioneer* spacecraft is designed to tell 'scientifically educated inhabitants of some other star system' who launched the craft. Without prior knowledge of our use of symbols, however, the inhabitants would not be able to decipher the message. (Courtesy of N.A.S.A.)

This type of project is not something in which we can participate at present, though we could take steps to ascertain whether or not alien peoples have attempted to do something along these lines with respect to our own family of planets. Naturally this assumes the existence of races with technological prowess far in advance of our own. Such a premise we are almost bound to accept, for if life proliferates within our galaxy a proportion of it at least must come within this category. Such superior communities, therefore, might be both able and willing to dispatch probing, unmanned vehicles to the planetary regions of other suns. Professor R. N. Bracewell of Stanford University believes that such exploration might be 'intense' in some of the stellar systems lying 'close' to our sun. In other words, if really advanced civilizations exist in the proximities of Alpha Centauri, Tau Ceti, Epsilon Eridani, and Epsilon Indi, then our sun and its system of planets might be an obvious target for highly instrumented probes. Bracewell in fact goes on to speak of such peoples 'spraying' a number of suitable stars, probably as many as a thousand, with objects of this nature. He further envisages each probe being sent into a circular orbit about one of the selected stars at a distance corresponding to the mid-point of that star's habitable zone. These would be armoured against meteorites, shielded from damaging radiation, and provided with automatic radio equipment with which to attract the attention of advanced technologies on nearby planets. The idea that even now inanimate emissaries of a remote alien civilization may be in our midst is highly intriguing. Does the impalpable darkness of space surrounding our small planet already contain an object that has 'seen' dawn break on a world whose sun is not ours; that has been fashioned by hands that are not our hands; that has 'heard' the strange cadences of unknown tongues?

This undoubtedly is reminiscent of science fiction, yet it should not be dismissed merely on that account. We are not trying to suggest that alien robot ships are buzzing round our planet or that the peaceful star-strewn night sky contains some hidden, nameless menace. It is all too easy to link such thoughts with flying saucers and thereby draw wholly unwarranted conclusions. We do say, however, that the arrival within the solar system of such an object is *not* beyond the bounds of possibility!

That the sending of a radio probe would have certain advantages over a normally transmitted signal is indisputable. The signal would be stronger since it would not have suffered attenuation by virtue of immense distance. Being stronger it would be considerably easier to detect. Moreover, the presence of such a probe within our system would obviate much of the difficulty concerning frequency choice and so on.

Bracewell suggests that even now we should start scrutinizing the spaces within the solar system for possible probes injected into it by technologies located around some of our nearer stellar neighbours. By so doing we would, in a sense, be paying a modicum of attention to *all* stars potentially capable of contacting us.

Later in this chapter we will briefly examine the question of probable alien visitations that may at one time or another have been made to our planet. The legend and folklore of many lands contain tales of peculiar objects in the sky, of strange lights, and other odd manifestations. Is it possible that a few of these old fables and superstitions have some basis in fact? The trouble with legends, of course, is precisely that which renders them legend – age! As year follows year, as generation succeeds generation, the record becomes increasingly distorted. Moreover, those who lived centuries ago tended, hardly surprisingly, to ascribe supernatural causes to perfectly natural phenomena.

The years following the Second World War are full of tales of flying saucers. Much has been said and written of all this space-borne crockery, a high proportion of it the most arrant nonsense. Obviously a large percentage of so-called sightings can be dismissed out of hand. Many undoubtedly owe their origin to meteorological balloons, high-flying aircraft, peculiar cloud formations, and too vivid (perhaps too eager!) imaginations. A few, however, are considerably more difficult to explain rationally.

Whether or not U.F.O.s (Unidentified Flying Objects) should be considered in the role of possible interstellar radio probes is a moot point. A probe would after all seem more likely to remain aloof from the atmospheric regions of a planet.

Bracewell considers that a *pilot* probe might first be sent. This would 'listen' for evidence of advanced technologies and report back accordingly, either directly or via relay probes. If this report should be positive – and in the case of Earth it presumably would – then we might be able to look forward to the arrival in due course of a 'more sophisticated mission', a prospect calculated to arouse either interest or foreboding.

The idea of a probe programmed with an extensive store of information and containing a computer by which it could 'converse' with us is a logical extension of the concept. As we have said, it is just possible that such a probe is already lurking somewhere within the boundaries of the solar system. Here Bracewell raises a most interesting point, suggesting that in order to ensure the use of a frequency that could penetrate our atmosphere yet lie in a band certain to be in use, the probe would first pick up some of our domestic radio signals, then re-transmit them back to Earth. This would lead to *radio echoes having delays of seconds, even of minutes*. These words are peculiarly significant in the light of an event that took place over forty years ago. In the weekly scientific journal *Nature* there appears in the issue of 3 November 1928 a letter written by one Jørgen Hals, a radio engineer of Bygodø, Oslo, to physicist Carl Størmer:

At the end of the summer of 1927 I repeatedly heard signals from the Dutch short-wave transmitting station PCJJ at Eindhoven. At the same time as I heard these I also heard echoes. I heard the usual echo which goes round the Earth with an interval of about 1/7th of a second as well as a weaker echo about three seconds after the principal echo had gone. When the principal signal was especially strong, I suppose that the amplitude for the last echo three seconds later, lay between 1/10 and 1/20 of the principal signal in strength. From where this echo comes I cannot say for the present, I can only confirm that I really heard it.

Størmer initiated certain tests as a result of this communication and on 11 October 1928 these bore some fruit. During the afternoon of that day Station PCJJ in Eindhoven emitted very strong signals on 31·4 metres. Both Hals and Størmer heard very distinct echoes several times, the interval between signal and echo varying between 3 and 5 seconds, most of them coming back about 8 seconds after the principal signal. Sometimes two echoes were heard with an interval of about 4 seconds. Physicist Van der Pol confirmed these observations in a telegram that read: 'Last night special emission gave echoes here varying between 3 and 15 seconds. 50% of echoes heard after 8 seconds!' At the time these peculiarly long echoes were attributed by Størmer to auroral causes but the feeling today is that they have never been adequately explained. Six years later, in 1934, radio echoes of a similar kind from Holland were also heard. No one, least of all the writer, is trying to suggest that an alien radio probe from the stars was definitely responsible for these inexplicable effects. What has been given are simply the facts as they are known. The reader must endeavour to draw his or her own conclusions.

We can hardly quote the case of station PCJJ without also mentioning the very mysterious business of station KLEE. For the record, station KLEE is (or was) a television station in Houston, Texas. In July 1950 station KLEE changed its call-sign to KPRC-TV. During the autumn–winter period of 1953–4 Paul Huhndorff, chief engineer of KPRC, received letters from various places (notably England) reporting reception of the KLEE call-sign. One of these was from a firm in Lancaster who had picked up the call several times between September 1953 and January 1954 and had, moreover, photographs to prove it!

For a time the mystery died down but in November 1955 one of the electronics specialists living in Morecambe who had seen the original signals and whose set was equipped to pick up long-range signals, switched on and got the KLEE call-sign again!

One so-called 'explanation' put forward by an American girl journalist is that a certain anonymous inventor had faked the signal by transmitting a copy of the station identity card. Engineer Huhndorff of station KPRC is, however, on record as saying that the photos sent to him show the *actual* call-sign card used by station KLEE. This raises the question of how the anonymous inventor got hold of it, not to mention the amount of money that would have to be spent perpetuating a rather pointless and illegal practical joke.

In February 1962 the KLEE signal showed up yet again. This particular report ought perhaps to be treated with a certain measure of caution. However the facts, for what they are worth, are as follows. A viewer during that month tuned in and got the KLEE station call-sign. After this had flickered on and off a few times a picture appeared which showed a girl being chased by a man who caught up with her. The sound did not come through for a while but when it did this, according to the viewer, was the dialogue:

Girl: Let me go! You can't stop me – I'm going to tell the world what you plan to do.

Man and girl are now joined by another man of whom we see the head and shoulders (rear view) only. Jumbled dialogue follows between all three, then:

> Second man: My God, man, you can't do that. It will be a catastrophe!

Picture fades out and KLEE signal comes on again, superimposed by the single word 'Help' which flashes on and off for a while before the screen goes blank.

The viewer then phoned her local TV companies to find out which of them was broadcasting (or trying to broadcast) this tantalizing programme. None of them had the faintest idea of what she was talking about. We feel bound to repeat, however, that this part of the tale should be treated with reserve.

The chief engineer of KPRC (late KLEE) was as baffled as anyone else over these appearances of his station's now defunct call-sign. The chance of the KLEE signal hitting a celestial object and bouncing back is possible (if not probable) once but hardly so many times. And why should just one station be affected?

At this point we must add a further twist to the mystery. This concerns the electronics engineer of a small radio station in a town in Alabama who had a most inexplicable and rather unnerving experience. He was also an amateur radio enthusiast. The month was November. The engineer put out a general call and got an answer from a fellow amateur in Tokyo, Japan. After a couple of minutes' chat they signed off in the usual manner giving their respective call-signs. The engineer then put out another general call – and promptly received back the last fifteen seconds of his previous conversation!

A fifteen-second 'bounce-back' would place the reflector some *considerable distance* away – unless his conversation was being played back to him. The possibly significant feature here is the fact that the call-sign of the Japanese amateur was JA1*PCY*. This bears a marked phonetic resemblance to *PC*JJ. The call was put out on a bearing of 315° at 500 watts on 21 mc/s (15 metre band).

It should also be added that the mysterious KLEE 'phantom' signals nearly always occurred over the autumn–winter period when the Northern Hemisphere is tilted away from the sun.

There are no records of the signal having been received anywhere south of the equator.

No distinct claims can be made for all these events but it would certainly seem within the bounds of possibility that something has been injected into our solar system which is capable of selective reception and re-transmission of some of our indigenous radio and television signals.

It is difficult to assess what form an extrasolar probe might assume. In some respects it might not be greatly dissimilar from the objects our race sends to the environs of some of our sister worlds around the sun, but in most it would obviously differ considerably. Here, after all, would be the product of a superior galactic community, capable of traversing interstellar distance. It seems reasonable to assume that it would avail itself, so far as power is concerned, of the light of the star to which it had been assigned. Indeed in the beginning it might be 'locked-on' electronically to this star that would thus guide it through the dark regions of interstellar space and eventually beckon it in.

Such a probe could re-transmit terrestrial radio signals not only to Earth but back to its home planet either directly or via conveniently spaced relay stations. So far as purely *radio* signals are concerned this would tell an alien community only that a reasonably advanced race occupied a planet orbiting the sun, for here again, of course, the language barrier presents itself. Ability to relay back television transmissions from Earth would obviously tell the aliens a great deal concerning ourselves and our civilization and to a sufficiently advanced race this might represent a feasible proposition. Here we feel bound to utter the pious hope that they would be suitably impressed by the things they might see!

How could the probe attract our attention, other than by tearing noisily across the skies of five continents and giving flying-saucer addicts their best field day yet? Re-transmission to us of some of our own radio signals as delayed echoes is, as we have already seen, a possibility. To our way of thinking this may not seem to be a particularly appropriate method, but as we shall see later the logic inherent in minds light years distant might be very different from our own.

The transmission of alien voices to us by the probe might attract our attention but would certainly tell us little. Scenes or actions on its home planet televised to Earth would, were it possible, be of immense interest. This, however, represents wishful thinking on our part. Simplicity, it would seem, should always be the keynote, especially with regard to initial contacts. A probe that could transmit the image of a stellar constellation indicating at the same time the star of its origin is a fairly obvious possibility. There is, of course, a difficulty here. To the inhabitants of an extrasolar world the constellations might appear in rather different guises. After all, these are merely convenient star groupings whose constituents have for the most part no physical relationship. Indeed the distances separating us from the various stars in any one constellation generally differ very widely indeed. Since such star configurations are, therefore, dependent only on the position of the viewer, the constellations seen from the skies of Earth will not be quite those appearing in the skies of Epsilon Indi III, which in turn will not be those in the skies of Tau Ceti V. However, a race competent to hurl an instrumented automatic probe across light years of space should be able to predict and project the constellation patterns likely to be seen from the planet whose destination it is. (See Figure 27b.)

We must remember that messages from alien probes travelling back with the speed of light towards their home stars are still subject to time-lag. If such a probe (or indeed probes) has entered the solar system and if its sponsors are by some means able to render intelligible the transmissions from Earth's radio stations, then the state of these people's knowledge concerning the affairs of our planet is incomplete. If, for example, the probe came from a world of Alpha Centauri or Tau Ceti then its sponsors must be considered reasonably up to date with terrestrial history. If, however, it originated in more remote regions the position is quite different. People near Vega, for example, are presently being regaled by the words of Adolf Hitler and his friends, those in the environs of Arcturus by news of worldwide economic collapse, while these who acknowledge Aldebaran as their sun are learning of the well-organized and effective mass slaughter of the First World War.

The question of probes being sent into the solar system by races on planets orbiting certain of the nearer stars leads fairly

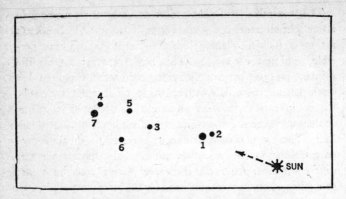

Figure 27b. The changing aspect of the constellations.

How the constellation of Orion would look from a certain region in interstellar space far removed from the solar system. Note how the configuration has changed out of all recognition and the great differences in distance between the respective stars and the sun.

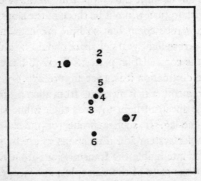

The beautiful and well-known constellation of Orion as seen from Earth and the solar system.

naturally to one that is even more intriguing – the possibility that in the past intelligent beings from some of the star worlds visited Earth. This is admittedly something of a digression from our main theme. Since, however, we are discussing communication between ourselves and alien intelligence it is probably more an extension than a digression.

To concede the possibility of such visitations either in the recent or remote past means conceding also the fact that other civilizations have solved the tremendous problems inherent in interstar travel. At present we ourselves remain acutely aware of the fearful magnitude of this task, of the terrifying time and distance barrier involved. That we can as yet see no way round this does not mean that a way will never be found. It is essential to realize that despite our recently accelerated technology and our first faltering footsteps into space, we are only paddling among the shallowest fringe waters. The fact that *we* have as yet found no answer does not render an idea incapable of realization. To aboriginal tribes our modern achievements must seem like miracles. How, therefore, should we contemplate the probable achievements of people who may well be our elders by many thousands of years, people to whom *we* are the backward tribe?

It is easy and less than fair to accept the possibility of interstellar travel by attributing its accomplishment to advanced alien technology and leaving it at that. This gives no indication as to *how* the seeming miracle may have been achieved.

The use of cryogenics is one of the first ideas to suggest itself. Reduced to its simplest terms this means induced hibernation for the occupants and crew of a star vessel by a form of deep-freeze treatment. Shortly before the fully automated starship reaches its destination these people are awakened and revived. To us no realization of this concept is in sight though the first possibilities are beginning to suggest themselves. That civilizations greatly in advance of our own could have perfected such schemes seems plausible. This idea, that the star-powdered night sky may contain somewhere in its depths strange space vessels containing inert living aliens, is almost as macabre as it is fascinating.

To writers of science fiction the idea of hyper-optical velocities has always been an appealing one – indeed the only possible one if hero and heroine are to conduct their affairs, adventurous and otherwise, on a remote alien world. Unfortunately Einstein exerts a very inhibiting influence when it comes to translating such dreams into reality. The theory of relativity tells us that the mass of a body increases in direct proportion to its velocity and if moving with the speed of light it would have infinite mass. This seems as good a way as any other of saying that travel at light velocity is impossible.

Could it be there is a way round this seemingly impossible barrier – a way already found by certain advanced star peoples? It seems there just *might* be. It has been shown mathematically by Gerald Feinberg that there may be a counterpart to Einsteinian mass – one represented by particles moving with infinite velocity. As these approach the speed of light they are in fact *slowing down*. Such particles have been given the name tachyons and according to Feinberg must be a *billion* times faster than light! Reduced to sub-light velocities they simply cease to exist. Unfortunately tachyons are at present merely a mathematical abstraction whose actual existence cannot yet be proven. Attempts to secure this proof are currently proceeding.

A recent experiment involved the use of a hollow sphere made of lead, the interior of which constituted a near perfect vacuum. Elements contained therein were bombarded by high-energy particles. According to theory any tachyons produced should travel across the sphere at velocities in excess of that of light and by so doing give rise to a phenomenon known as Cerenkov radiation, which takes the form of a distinct bluish glow. This, it is considered, can only happen when particles are travelling in a particular medium with a velocity greater than that of light. In a vacuum only tachyons, it is postulated, can travel with such velocities. In the experiment Cerenkov radiation was not apparent, from which it could be deduced that under the conditions then prevailing tachyons were not present.

Perhaps the result of this particular experiment should be regarded as indecisive rather than negative for it may well be that experiments of a much more sophisticated and elaborate design will be necessary to prove categorically the existence of such elusive particles as tachyons. Current scientific opinion tends to believe on reasonably strong mathematical grounds that tachyons probably do exist and that sooner or later some measure of proof will be forthcoming. Clearly the design of an experiment to establish this existence can be neither a simple nor a straightforward matter. The search is, after all, for infinitesimally small subatomic particles that can only exist by moving at velocities *in excess* of 186,000 miles per second.

Particles exceeding the speed of light, it must be added, are not in fact contravening the basic Einsteinian concept that sees light velocity as a limiting factor. The explanation for this seeming anomaly involves what is known in nuclear physics as rest mass. A digression into these realms would, however, be somewhat out of place in these pages.

How might aliens make use of tachyons? Perhaps by harnessing them or creating them by artificial means. A space vehicle of appropriate design could then be accelerated to the speed of light, perhaps using a form of photon-propulsion unit. This point reached, a tachyon-drive unit could then take over, enabling velocities thousands of times that of light to be attained. One obvious point of difficulty is the fact that the actual speed of light would have to be achieved before the tachyon drive could become effective. Though capable of velocities very close to that of light a photon drive would still not be able to attain that of light itself. Our alien friends, it would seem, must somehow adapt their technique to bridge this vital gap.

Velocities of such a tremendous order would obviously present grave and not necessarily foreseeable risks. They can thus be divided into two categories – the known and the unknown. Among the former we must think immediately of the disastrous consequences of collision with the smallest speck of matter. Some form of repulsive or disintegrating force ranging ahead of the spacecraft would seem imperative. Neither can we ignore the aspect of time dilation and all its peculiar manifestations in these particularly bizarre circumstances. Of the unknown we can say little or nothing. The effects could and probably would be very peculiar indeed. It is just conceivable that the result of hyperoptical velocities of this order might tie in with the idea of travel through a different form of space, that is, an as yet unsuspected dimension.

This unsuspected fifth dimension is at present something of a paradox and indeed may be nothing more. Astronomical research of recent years has shown space to be more than a little unusual in a number of respects. These discoveries are not necessarily related to speculative extrapolations of the 'unknown dimension' type. Nevertheless, space is not apparently what it seems, and the concept of a simple unending 'nothingness' has taken more than a few severe knocks in the past decade or so.

One manifestation that springs immediately to mind is the existence of the celebrated 'black holes' – regions of space into which a star or a number of stars fell and from which no light or matter can escape. The fate of the original dying star or stars that collapsed to create the black hole is bizarre in the extreme for its substance is squeezed to infinite density at the centre of the black hole. It is to all intents and purposes crushed out of existence. In the process the vacuum of space–time outside the body becomes infinitely curved. To anyone so unfortunate or so foolish as to enter this region the effects would be catastrophic, for his or her body would be subject to a tidal force that would reach infinity. This hardly seems a medium suitable for spacecraft, and we would be the last to suggest that it was. However, one paramount truth emerges. The outer bounds of space could be very peculiar indeed. In truth all is not as it seems. Conventional ideas could be far removed from stark fact, and a very open mind should therefore be kept.

Conventional space is regarded by some cosmologists as being curved. In other words travel between the sun and Alpha Centauri, although to our senses a straight line, is in fact a curve. (See Figure 28.)

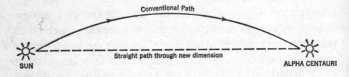

Figure 28

If instead we could puncture the 'bubble' of curved space and take the path shown by the straight dotted line this would represent a short cut. Since it involves transit through another dimension (sometimes termed non-space) the period of transit might be ludicrously short, indeed instantaneous.

We can point only to possibilities. We cannot as yet say that tachyons do exist, and, of course, the idea of an unknown dimension is even more academic. It may be, however, that space, matter, and time are entities not wholly in accord with our conventional beliefs. By their ability to take advantage of such possibilities cosmic civilizations several millennia our elders may have the power to roam our local star cluster with a high degree of freedom or at least direct their probes to selected points within it.

Neither can we discount progress made by advanced cultures in the realms of biology. Really long space voyages by our kind are rendered void by the fact that the human life-span is simply too short and that any prolongation we can presently visualize is totally insufficient. But even with us the possibility, indeed, the fact of organ transplant has arrived on the scene. So far success has been limited, if indeed it can be termed success. Nevertheless, the potential is unmistakable. We must look forward to a day when organ banks will be able to supply either rehabilitated natural organs or those of an artificial kind, when the human lifespan may as a result be doubled or tripled. Is, then, the idea of great stellar argosies plying through the galaxy on Magellan-like voyages so ridiculous? Explorers on Earth centuries ago spent months, even years, at sea. Conditions were primitive and often cruel, yet faith, courage, and resolution won through. Space exploration will eventually show a parallel. To some alien peoples it may already have done so. Perhaps our sun and its family of attendant worlds will yet prove to be a cosmic 'Indies' to a race of unknown form.

From the outset it must be made clear that consideration of this question is not synonymous with an inquest on the vexed and already grievously over-discussed subject of flying saucers. Some aspects may be valid, but it is more desirable to search the past, the remote past, and even legend, for instances that might be interpreted as alien visitations. It is absolutely essential to stress, however, that in no instance is positive contact with an extraterrestrial race inferred. We state only the facts, present such evidence as there is, and from this assess, so far as it is practical, the possibilities.

It should be emphasized that the concept of alien visitations to our globe in the remote past is not merely the prerogative of wishful thinkers. The idea has been endorsed as a possibility by several eminent figures in the world of science. Einstein, for example, is known to have been in sympathy with it. So is Hermann Oberth, one of the founders of rocketry. The Russian astrophysicist I. S. Shklovsky thinks that Earth may have been visited on at least one occasion, as does the American spacebiologist Carl Sagan.

A factor that may tend to add a modicum of credibility (and respectability) is the relatively 'sudden' appearance of human intelligence. The remains of the Neanderthal man, on the best available evidence, are some 100,000 years old. This is young indeed compared to the relic found near Cairo – a lower jaw attributed to the Oligocene Age and therefore between 30 and 40 million years old. Fossil homonids have also been found in Australia, Borneo, and Africa, proving that creatures akin to man have existed for a very long time. It would, perhaps, be more accurate to say that transition from homonid to true man proceeded gradually over many millions of years, yet the acquisition of intelligence, the ability to reason, took place with a relative suddenness that is surprising. Intelligent man seems to have appeared about 40 to 50 thousand years ago. Crude weapons were devised and even cruder stone tools. He learned to paint on cave walls, and he became able to make use of fire. Was this 'sudden' gift a natural acquisition – or did he pick it up from representatives of an alien race? These are only passing thoughts, probably irrelevant, yet they keep returning with a curious insistence.

If we assume that visits to this planet have in the past been made by alien beings it is not unreasonable to query the notable lack of clear evidence. On the other hand we must remember that if such visits were few in number and of short duration – and this seems a readily acceptable premise – then quite obviously material remains will be negligible. Moreover, if visitations took place in early geological ages then subsequent changes in the earth's crust must almost certainly have resulted in complete elimination or permanent concealment of all evidence. Indeed a crippled starship might have been encapsulated for all time by the tortured, wracked crust of our then primeval planet or encased for ever in a magmatic tomb miles deep by a mighty display of prehistoric vulcanism.

Probably the most interesting and appropriate place to start our search is in the Old Testament of the Bible, for here there are one or two instances of singular note. In the Book of Ezekiel, for instance, we find the following:

> And I looked and behold a whirlwind came out of the north, a great cloud and a fire unfolding itself, and a brightness was about it, and out of the midst thereof as the colour of amber, out of the midst of the fire.
> Also out of the midst thereof came the likeness of four living creatures. And this was their appearance, they had the likeness of a man.
> And every one had four faces and every one had four wings.
> And their feet were straight feet and the sole of their feet was like the sole of a calf's foot; and they sparkled like the colour of burnished brass.

And they had the hands of a man under their wings on their four sides, and they four had their faces and their wings.

Their wings were joined one to another; they turned not when they went; they went every one straight forward.

As for the likeness of their faces they four had the face of a man and the face of a lion on the right side and they four had the face of an ox on the left side; they four also had the face of an eagle.

Thus were their faces; and their wings were stretched upward; two wings of every man were joined one to another and two covered their bodies...

As for the likeness of the living creatures, their appearance was like burning coals of fire and like the appearance of lamps; it went up and down among the living creatures and the fire was bright and out of the fire went forth lightning.

(Ezekiel 1:4–13)

Now this is a very colourful and imaginative description, and due allowance must presumably be made for inaccuracies in translation. Nevertheless, here quite obviously is a description of living beings who were certainly not human. Particularly significant are the passages that speak of a 'whirlwind come out of the north, a great cloud and a fire unfolding itself'. Suppose for a moment a space vessel were to land near us and that hitherto we had no knowledge of such things and no modern technology. Would we not tend to describe the event in roughly similar terms? Suppose too that beings of different form were to emerge. There would be excitement, confusion, and fear. Descriptions would vary widely. It is safe to assume that few would be so imprudent as to venture close to the beings. Descriptions of the emerging creatures would be vague with hysteria and imagination given free rein. Their appearance would almost certainly be likened to that of legendary creatures, for this is about all that an early and superstitious people could draw upon. The description of the creatures then is significant not so much by virtue of what is said (although by any standards this is of great interest) but simply because it portrays beings so obviously and so strikingly different.

The passage becomes even more intriguing as it continues:

Now as I beheld the living creatures, behold one wheel upon the earth by the living creatures with his four faces.

The appearance of the wheels and their work was like unto the colour of a beryl; and they had one likeness; and their appearance and their work was as it were a wheel in the middle of a wheel...

As for their rings, they were so high they were dreadful; and their rings were full of eyes round about them four.

And when the living creatures went, the wheels went by them; and when the living creatures were lifted up from the earth, the wheels were lifted up.

(Ezekiel 1:15–19)

Here might seem the description of a powered vehicle couched in terms we would expect from an ancient and entirely non-technical civilization. The reference to 'the wheels and their work' being 'like unto the colour of a beryl' is intriguing. A beryl is a pale-green, yellow, or light-blue stone and is composed of a silicate of beryllium and aluminium. Beryllium itself is a very light silvery-white metallic element that when alloyed with copper produces an extremely hard metal used in the production of high-grade gear wheels. With aluminium, a hard and corrosion-resistant alloy is obtained. Beryllium on its own is extremely useful in the manufacture of high-speed aircraft and missiles because it is both stronger and lighter than aluminium.

Now the phrase used in Ezekiel, 'like unto the colour of a beryl', need not be taken too literally. If the object were the product of an extraterrestrial technology the words could easily relate to an aluminium alloy with or without beryllium. We must bear in mind that the people of Ezekiel's day had never seen aluminium, and so likening it to a material (beryl) that is pale green or light blue in colour is not really so ridiculous. Remember again the prevailing confusion and fear, the effect of distance, and perhaps also of the reflected light of an intensely blue sky.

The passage goes on, 'And when they went I heard the noise of their wings like the noise of great waters . . . as the noise of an host.' To anyone who has heard the thunderous roar of a departing rocket this description will seem singularly appropriate.

Now it is all too easy (and tempting!) to attribute Ezekiel's vision or experience to a terrestrial visitation by superior space beings. Nevertheless the description in many aspects fits the arrival of a space vessel.

Ezekiel lived in the sixth century B.C., and was one of the major prophets deported to Babylonia by Nebuchadnezzar in 597 B.C. These were times and places far removed from our present-day world, yet here is a description of machines and creatures strangely applicable to the advent of alien beings bearing the products of a supertechnology – all so remote from the heat, dust, fears, and superstitions of the Middle East of 2,500 years ago.

Of interest, too, in this context is the famous account of the destruction of Sodom and Gomorrah, the notorious twin cities.

Then the Lord rained upon Sodom and upon Gomorrah brimstone and fire from the Lord out of Heaven.
And he overthrew those cities and all the plain and all the inhabitants of the cities and that which grew upon the ground ...
And he [Abraham] looked toward Sodom and Gomorrah and toward all the land of the plain, and behold, and lo the smoke of the country went up as the smoke of a furnace.

(Genesis 19:24–58)

The phrase 'fire and brimstone' is not uncommon in the Bible. It also has connotations of volcanic action, but it seems extremely unlikely that any form of vulcanism took place in this region of the world at that particular time. An earthquake or even the fall of a great meteor is feasible although the latter would have been expected to leave a vast crater visible to this day. Just what agency, then, did God use to destroy the infamous cities? Fairly recently a Russian scientist, Kasantev, has suggested that their destruction might be attributed to a nuclear explosion when extraterrestrial creatures disposed of excess nuclear fuel before taking their departure. The inhabitants of the cities were warned to evacuate the region, not to remain in the open, and above all to refrain from looking. The unfortunate Lot's wife, as we all know, did look and was as a consequence transformed into a 'pillar of salt' (Gen. 19:26). This incident is rather difficult to explain but we might suggest that Lot's unwitting spouse did not so much look back as linger some appreciable distance behind her more sensible husband.

'But his [Lot's] wife looked back from *behind* him and she became a pillar of salt.' Whatever happened to this luckless and now legendary woman, it seems highly improbable that she was actually transformed into a block of sodium chloride! The reference to 'salt' must not be taken literally. If the destruction of the cities is attributable to nuclear blast then it is not unlikely that the woman suffered the searing effects of heat radiation.

The region is known to have been prone to seismic disturbances, but it is difficult to see how an earthquake could have produced the effects described. It could be argued that a present-day excavation of the spot where Sodom and Gomorrah once stood might, in the light of modern knowledge, reveal fresh information on the mysterious and sudden destruction of the twin cities. Unfortunately the exact location of these cities is not known, and it may be that their remains have been covered for ever by the waters of the Dead Sea.

Another biblical episode regarded by some researchers as worthy of closer scrutiny is the famed and catastrophic collapse of the walls of Jericho. In Joshua 6:20 we find the following passage: 'And it came to pass when the people heard the sound of the trumpet, and the people shouted with a great shout, that the wall fell down flat, so that the people went up into the city, every man straight before him, and they took the city.'

The walls in question are reputed to have been about 21 feet thick. We can be quite certain, therefore, that whatever agency led to their dramatic collapse it was *not* the combined assault of massed trumpets and human voices. We may accept that on occasion a singer or violinist may shatter a fine crystal vase by a note of appropriate pitch but this must surely be regarded as the limit.

Why did the walls of Jericho collapse so catastrophically and at such a militarily convenient moment? Assuming that no seismic disturbance had taken place or any prior undermining of the walls we must inquire whether or not some sophisticated acoustic device was responsible.

Vladimir Gavreau of the Electro-Acoustics Research Institute in Marseilles built in 1964 what has come to be described as a 'sonic gun'. Essentially this was a contraption comprising sixty or so tubes through which compressed air was passed. An audio frequency of 196 cycles per second resulted, the effects of which were profoundly disturbing. Walls cracked and certain internal organs vibrated in a most painful fashion. Intrigued by what he had achieved Gavreau built a much more powerful device, of 2,000 watts and 37 cycles per second. Fears that buildings over a radius of several miles might collapse prevented its being tested at full strength. Recent reports from this quarter tell of experiments with a lethal sonic gun that is 75 feet in length and operates on a frequency of 3·5 cycles per second. Such a device operating on a low frequency of this order would certainly have reduced the walls of Jericho – and presumably much of Jericho as well!

Where, however, could the forces investing this city have procured such a sophisticated device? It could hardly have been the product of their technology (such as it was) or that of any other contemporary terrestrial people. *If* it existed it came from elsewhere. That elsewhere could only have been extraterrestrial!

So much for biblical aspects of the problem. Let us now look into other quarters. Perhaps as good a starting-point as any is the case of the great enigmatic stone spheres of Central America, notably those of Costa Rica. Most of these are of lava, granite, or some other form of very hard igneous rock, and their dimensions are truly remarkable. One having a diameter of 7 feet 1 inch has been placed in the grounds of a building in San José, Costa Rica, the capital.

Many mysteries surround these peculiar objects and are summarized as follows:

1. Since there are no quarries or other sources of the particular rock in the vicinity they can only have been transported to their present locality.

2. In some cases these objects have been found amid dense rain forest and primeval jungle. Evidence points to the fact that they were there *before* the jungle!

3. The balls are perfectly spherical with smoothly polished surfaces and could not possibly owe their origins to natural causes.

4. In other cases they have been transported to mountain tops. Marvellous as were the civilizations of the ancient Aztecs or Mayas we must regard this as a feat beyond their powers.

The smallest of these objects have diameters of only a few inches. At the other end of the scale some are over 8 feet in diameter and weigh as many tons! In a number of instances the distribution pattern is odd – almost as if their creators had been endeavouring to impart a message. While there is no obvious indication that these things *are* the work of aliens it is very difficult to attribute them to indigenous peoples of the region. Production and transportation of such objects in our present day and age would represent neither an easy nor a straightforward task.

South from Costa Rica into the ancient land of the Incas we find another fascinating and perplexing relic – the odd markings on the plain of Nazca. A true impression of these can only be obtained from the air. Essentially they are great broad strips running for miles and to a large extent in parallel. There are also numerous intersections and on the 'islands' of normal terrain among the strips the outlines of peculiar animal-like figures can be discerned.

From ground level the strips are merely broad furrows that reveal the pale yellow subsoil beneath the brown sand of this near-desert region. How, it might well be asked, have these survived the passing of centuries? The answer is not hard to find – lack of rain. The climate is arid and very hot. Rain, if it comes at all, is infinitesimal in amount (one report quotes a figure of 20 *minutes*' precipitation *per year*!). The only weathering agent is the wind, but this is rendered impotent by virtue of a peculiar protective 'skin', presumably an oxidative coating, that covers the strips.

Who was responsible for these markings and why were they made? From the air the entire system is highly reminiscent of a large modern airport with its complex of runways, intersections, and taxi-ing strips. So systematic and symmetrical does the system appear that the feeling that it must have been planned and drawn originally on a small scale is hard to dispel. If this is so, then a high standard of surveying must have been necessary to translate such a system into reality. From the peoples of this region in the remote past this seems most improbable. What was the purpose of this feature? Landing ground? Pointer to the seasonal rise and setting of certain stars? Is it indigenous – or alien?

A language closely akin to Magyar is known to have been spoken in and around the Quito region of what is now Ecuador long before the arrival of the Spanish. How did a tongue of central Europe come to be spoken here (or vice versa)? Distance and a great ocean separated these areas yet we find family and place names that are identical. To the best of our knowledge there was not, could not, have been contact between the two.

There were no fast forms of transport, no electronic means of communication – *or were there?*

Burial customs, too, are near-identical. One feature in this respect is particularly noteworthy. The Magyar funeral oration, it is said, ended with the words, 'He will vanish into the stars of the Great Bear.' Oddly enough in certain South American valleys are burial mounds laid out precisely according to the stars of this famous and legendary constellation. Coincidence? Perhaps. Should the thought be dismissed entirely? To this question each individual must apply his or her own answer.

Certain features in the Libyan Desert could also, it is seriously suggested, be due to alien visitations. One of these, a platform of large stone slabs, has been seen as a form of launching ramp for the ship's return journey or as some kind of commemorative memorial raised by the aliens. While both ideas are remotely possible it does seem safer to attribute such works to an earlier terrestrial society. Nevertheless a closer scrutiny of the Baalbeck Terrace, as it is known, might be worthwhile.

Even less easy to explain is a peculiar object found in an early Egyptian tomb (3rd or 4th Dynasty) around the end of the nineteenth century. This was apparently a piece of polished copper that might conceivably have constituted part of a copper mirror. Eventually it found its way into a Cairo museum. As it lay there a shaft of sunlight happened to fall directly on it whereupon, to the intense surprise and interest of all concerned, a perfect spectrum was seen to be projected on to the ceiling. Closer examination of the object revealed that on its surface were inscribed a large number of exceedingly fine lines. The thing was in fact a diffraction grating. For the benefit of readers unfamiliar with the physics of light it should be explained that a diffraction grating is a thin plate, generally of transparent material, on which are ruled parallel straight lines, numbering about 15,000 to the inch. Without going deeply into physics it can be said that a beam of light falling on this produces the typical spectrum or 'rainbow-colours' effect.

Now despite the ability of the ancient Egyptian to work with metals there is little doubt that the production of a diffraction grating would have been beyond their powers. Even had they been able to produce such an object they would hardly have done so, for, of course, they had no conception of the nature of light.

If the object were produced on earth many centuries later, then we must ask how it found its way into a sealed and hitherto undiscovered tomb? To these questions there appear to be no rational answers.

Peculiar cave paintings exist both in Japan and in Russia, which depict anthropomorphic shapes in some kind of armour that bears more than a passing resemblance to spacesuits. It is risky, of course, to attempt to draw conclusions from this sort of thing. Many ancient suits of armour are vaguely like spacesuits (and vice versa)!

There are also instances in several other parts of the world of small statues in attire reminiscent of astronaut's garb and even paintings of objects that bear more than a passing resemblance to air- or space-craft. What are we to make of these and what conclusions, if any, dare we draw? To a large extent these representations tend towards caricature – crude attempts to portray what may have been actuality. To simple, primitive peoples, space vehicles and spacesuited figures would be utterly divorced from reality. In the circumstances what they would draw or portray might in a sense be akin to what a young child would reproduce today. Figure 29 and Figure 30 are drawings made by the writer's two daughters (Anne, aged thirteen, and Karen, aged ten). They were not copied and are in both cases purely a child's idea of what a spacecraft and spacesuited figure would look like. Admittedly the ideas of children must be coloured by what they see of contemporary space vehicles in the press and on television and probably to an even greater extent by the rather peculiar creations appearing in some of their comics. The balance in this comparison may be redressed to some extent by the fact that ancient drawings (if they *are* of space vehicles) were made by mature men of their day. There can be no strict comparison, but what results is interesting for there is certainly a likeness between what grown men of past ages having *no* prior knowledge of space vehicles *could* have reproduced and what today's child with a little knowledge *does*.

In 1919 a man by the name of Charles Fort published a rather remarkable work bearing the title *Book of the Damned*, in which he mentions several instances of possible extraterrestrial visitations. One of these concerns copper mines in the vicinity of Lake Superior, in Canada, that had apparently been worked exten-

Figure 29. (Anne)

Figure 30. (Karen)

sively and methodically long before their discovery by white settlers.

Another instance deals with a hard steel 'cube' found in a Tertiary coal deposit in Austria. This steel cube was embedded within a large piece of coal by workmen intent on using the coal as fuel. The object weighed 785 grams (about 1¾ lb.), and its dimensions were 67 millimetres by 67 millimetres by 47 millimetres. There was also a deep, straight incision that ran completely round it. The thing was originally deemed to be a meteorite, and as such it was placed in the Salzburg Museum. It is rather difficult, however, to accept this object as a meteorite. Certain metal sulphides are known to crystallize in the form of cubes but the dimensions of the Salzburg object would seem to render this most unlikely. Moreover, this leaves unexplained the enigmatic groove running round it. In nearly all respects the origin of this object would appear to be artificial. If the latter assumption is correct then its discovery amid a Tertiary coal deposit defies all rational explanation.

In November 1869 there was found during mining operations near Treasure City, Nevada, a piece of orthoclase feldspar in which was deeply embedded the imprint of a two-inch screw. The screw itself had long since rusted away. This particular rock was formed millions of years before the advent of terrestrial man. The piece in question was taken to the San Francisco Academy of Science, but so far no rational explanation has been offered for the existence of the impression.

This is not the only instance of nails and screws found in inexplicable places. A piece of gold-bearing quartz brought from California by a doctor was accidentally dropped. This split open to reveal a small, slightly corroded nail. In a quarry in northern England there was found protruding from a piece of stone a badly corroded nail. Approximately half of the nail was embedded within the stone while the remainder (and the stone) had been surrounded by a bed of glacial debris laid down during the Pleistocene period.

The record of small artificial objects found deep within the crust of the earth is considerable. Not only is it comprised of screws, nails, and geometrical metal cubes but also of instruments, small statues, peculiar coins, and so forth. A fairly recent report concerns the discovery in the mountains between Tibet and China of over 700 stone discs. Each disc possesses a hole at its centre from which irregular grooves spiral outwards towards the edge. These discs were not embedded in the ground but were found in caves by an archaeological expedition. Some were removed to Moscow for a detailed laboratory examination. It was subsequently claimed that traces of metal, notably cobalt, had been detected in the discs, and one report states that the discs vibrated when scraped as if they bore an electric charge. Oddly enough there is a legend in this remote region that tells of 'giant yellow-faced men who came down from the skies' many hundreds of years ago. In other caves in the region archaeologists have discovered graves reputed to be about 12,000 years old in which have been found the remains of human beings possessing huge skulls and small underdeveloped skeletons.

Other finds also include remains of peculiar beings. There is, for example, the case of the four prospectors working in the hills near Eureka, Nevada, in 1877 who came upon an unusual object protruding from a ledge of rock. Their curiosity aroused, one of them scaled the cliff to secure a better look. To his surprise the object turned out to be a human legbone broken off just above the knee. The prospectors carefully removed the bone from the rock. Eventually it was examined by doctors who stated that it was undoubtedly human. What was so disturbing, however, were its almost incredible dimensions. The length from knee to heel was three and one quarter feet; thus its owner in his day must have been approximately twelve feet tall!

An entire skeleton of these dimensions was uncovered by a party of troops engaged in digging a pit in California in 1833. It was surrounded by large blocks of porphyry (an igneous type of rock) in which strange symbols were inscribed. Further evidence that large biped creatures (indigenous or otherwise) have at one time or another walked on the face of this planet was afforded by the discovery as recently as 1926 of two 'human' molars in a coal mine in Montana. These lay in seams some thirty million years old and were three times as large as present-day teeth.

Again we cannot say with any assurance that such remains are the relics of visiting cosmic beings. It still seems more reasonable to suppose that before the dawn of history very large human beings peopled parts of our planet. And yet if this is so, should evidence of their one-time existence not be *more* widespread? The fact that very few remains have actually been found seems more indicative of brief visits!

There is on record an event within very recent times that has a bearing on giant beings. This, however, is unsubstantiated and would seem to be part of the contemporary U.F.O. scene. For what they are worth, the details are these. At dawn on 18 October 1963, the driver of a truck going between Monte Maiz and Isla Verda in Argentina saw a large circular metallic object approaching. The craft halted and from it emerged a number of very tall figures, reputedly thirteen feet tall. Just how much note should be taken of this report is a very moot point. In the circumstances, however, it is at least interesting.

The possibility that advanced alien forms of life visited our planet in remote ages long before the advent of man or indeed before the coming of life has raised the possibility in the minds of some that terrestrial life is not after all an indigenous entity. Clearly this is a very speculative matter, which in fairness cannot be disregarded. The noted science-fiction writer Arthur C. Clarke summed up the situation rather well when he said, 'Countless times in geological history strange ships may have drifted down through the skies of Earth.' If this is so, did one of these visitations lead to the seed of life on the hitherto sterile surface of our planet? Should this be the case, then man is not after all a man of Earth but a creature whose origins must be sought on the unknown planet of an alien sun. Is it possible, however remotely, that somewhere in the archives of a race light years removed from our own, there is to be found the enthralling story of how a great space vessel once descended through the tempestuous skies of a roaring primeval planet orbiting a golden-yellow star and, finding that world lifeless, barren, and hostile, proceeded on its lonely cosmic way?

Fairly recently an attempt has been made to explain the famous Siberian meteorite of 1908 as a violent detonation brought about when the nuclear motors of an alien starship exploded as it was descending. By and large the evidence is flimsy. A certain measure of credence and even respectability was accorded the suggestion by the Soviet scholar Kasantev. As a member of a post-Second World War expedition to the Tunguska region in Siberia, he was struck by the similarity of some of the destruction to the ruins of Hiroshima, which he had previously visited.

It is very tempting to disseminate spectacular ideas of this nature. Such an event *could* have occurred. Alien visitors to our world *may* have died suddenly and hideously in the catastrophic nuclear detonation of their great starship.

Kasantev has said, 'On that day our visitors from space met with a terrible end.' He goes on to ponder whether or not others will one day follow.

Whatever the cause, there is no doubt that the explosion was one of tremendous force. At Varanova, 40 miles to the south, ceilings collapsed and windows were blown out as thunder rumbled and muttered ominously in the north. Meanwhile, nearly 400 miles to the south, the famous Trans-Siberian Express was halted only a few feet from disaster when before the horrified eyes of its driver the track ahead heaved and buckled as if imbued with life.

It was not until 1927, nineteen years after the explosion, that the first scientific expedition reached the devastated area. The scene that met the members' eyes was bizarre. Trees had been blown down, and in some cases violently uprooted, over a radius of nearly 40 miles! Those which had been uprooted lay with their torn and shredded roots pointing towards the centre of the explosion area like the spokes of a great wheel. One witness, interviewed by the expedition, said: 'The fire came by and destroyed the forest, my reindeer and all the other animals.' Indeed out of a herd of nearly a thousand reindeer only the terribly burned and mutilated carcasses of a few were ever found. The rest had completely vanished, swept utterly out of existence. In all, an area measuring more than 20 square miles had been totally devastated. A few holes in the ground were found, but of any larger crater there was no sign. Borings made at the site of these small holes to a depth of nearly 30 feet into the frozen ground yielded no clue as to the cause of the cataclysmic blast. That some vast object from outer space had collided with earth seemed obvious enough, although the precise point of impact is difficult to determine.

Other inhabitants of the region when interviewed by scientists confirmed that the 'thing' had been *seen* in a cloudless sky over an area of nearly 1,000 square miles. According to observers near the scene, it 'had made even the light of the sun dark!' Fortunately the area was largely one of peat bog and forest, very sparsely populated and, so far as could be determined, no human lives were lost.

What precisely happened that June morning in 1908 will probably never be known. Was the thing a meteorite, the head of a comet, a chunk of anti-matter – or an exploding starship? Over sixty years have since passed. The signs of devastation have disappeared. The evidence for an exploding starship is flimsy, yet for all we know creatures of unknown form on an unknown world may still mourn and remember loved ones who did not return!

Over the years observations of brilliant meteors, fireballs, and related phenomena have been frequent. Were a few not quite what they seemed? Instances in which definite meteoric fragments were found are clearly above suspicion. The same is true of occasions when all indications are clearly and typically meteoric. It is on the unusual sort of 'meteoric' occurrence that suspicion falls. The majority of these will, despite certain unorthodox features, be genuinely meteoric. A handful may not!

A rather typical example occurred on 9 February 1913, when observers in Canada were startled to see a procession of 'glowing red fireballs' moving slowly across the sky in apparently level flight. Their altitude was estimated as being about 80 kilometres (50 miles), and the sound of their passage was likened to the 'roar of an express train'. The phenomenon was investigated at the time by C. A. Chant, a professor at Toronto University, who verified sightings from Saskatchewan in the west right across the Great Lakes to the eastern seaboard. Reports were later received from Bermuda and Brazil. In every case observers stressed the apparently *level flight* of the so-called fireballs. More recent investigations into this particular occurrence suggest that what was seen was the destruction of a small *natural* satellite of Earth, of a midget and hitherto unknown moon. No remains were ever found, and it was concluded that the main mass had plunged into the waters of the south Atlantic. This however is very much a case of surmise, and no really satisfactory explanation has ever been produced.

So vast has the literature on flying saucers become that it is virtually impossible to go into the matter here. In view of the spurious nature of so many of the supposed sightings, it is not really desirable that we should. It is by now abundantly clear that a high proportion of these are due simply to erroneous interpretation of natural phenomena. We think especially of odd cloud formations and, if at night, of the auroral feature known as noctilucent clouds. Man-made objects such as meteorological balloons and high-flying aircraft are no doubt also the cause of many reports. Some of the more sensational accounts are so obviously hoaxes or the work of publicity-seekers that they are best treated with the derision they deserve. Even well-intentioned reporters can inadvertently embellish the saucer legend. Entire works have appeared by writers who go so far as to claim they have flown in U.F.O.s and conversed with their occupants (generally in English). With due respect and tolerance it is hard to see the point of such exercises!

If we eliminate all sightings that are in any way doubtful (and this includes most!) we are left with a handful which are not easy to explain. Generally these emanate from persons mature in outlook and trained to observe – scientists, fliers, Air Force personnel, and so on.

In one or two instances spacecraft have been reported hovering in the vicinity of high-tension power cables. These sightings are fairly well substantiated and usually describe the objects as 'glowing'. Not surprisingly this has led to a belief that in some peculiar manner the strange vessels were 'recharging' themselves at the expense of the appropriate electricity authority. One has to proceed with caution here for a plausible explanation is not easy to find. Recently, however, it has been reported that under certain conditions luminous effects of this nature *can* be produced by power lines themselves.

There is a facet of the flying saucer paradox that is rarely if ever considered. Most would expect an alien starship to be recognizable as such. In accepting that it is the product of a technology considerably more advanced than our own we would anticipate a craft vastly different from any space vehicle yet produced or envisaged. If, however, it is the creation of a society several hundred millennia ahead of our own it must almost certainly embody features beyond our comprehension. It might, for example, be transparent – or very nearly so. It might even be able to pass through what is to us an unknown dimension. A hos of intriguing possibilities spring to mind. Perhaps the peculiar nature of sightings should be considered in this light.

The furtive nature of contemporary U.F.O.s is intriguing. If these are genuinely alien craft bearing representatives of an advanced race with all the impediments of a fantastic technology, one would assume they would descend from our skies openly. The rational answer, of course, is that U.F.O.s do not and never did exist other than as figments of the imagination. Despite this there are those with a niggling doubt – a feeling that a cultured and infinitely superior cosmic people might just be keeping us under surveillance. Since the late 1940s we have been in possession of dangerous nuclear toys. More recently a 'primitive' capability in space exploration has appeared. To a grand cosmic federation all this might seem faintly ominous; it might seem to constitute a situation worth watching.

If aliens did first arrive during the Earth's Palaeozoic age there might exist within their ancient archives a record of that visit – of the conditions on the third planet of a star we call the sun. Have they decided it is time now for another look, another assessment – a reappraisal after countless ages?

It is best to retain if possible a completely open mind on the subject of flying saucers since it is one of considerable scope and implication. The writer can report (for what it is worth) the one and only occasion in which he observed what might have been a U.F.O. This occurred on the evening of 3 September 1957, when at about 9.00 p.m (British Summer Time) a brilliant golden object was seen heading quickly across the perfectly clear sky in a north-westerly direction. It should be stressed that this was prior to the era of Sputniks and artificial earth satellites. There was no vapour trail such as aircraft are prone to leave under certain conditions. Neither was there any audible sound. The altitude of the object was considerable although impossible to assess with any degree of accuracy. Eventually the 'thing' disappeared over the northern tip of Arran, a large island lying off the south-west coast of Scotland. Seen through binoculars the object was undoubtedly circular and showed a sharp contour. No markings or features could be discerned. The facts of this incident can be verified by the wife of the writer who was present and also witnessed it.

That in our universe life abounds seems today a very likely premise. Other civilizations may be signalling to us, or sending instrumented probes in our direction. Some may in the past have visited us, employing a stealth so consummate that we were and are unaware of what has happened. Some even now may be preparing to do so. It would be ironic were we to find eventually that creatures from another solar system were more conversant with ours than we ourselves. Perhaps when our manned space vehicles reach Mars and beyond, their occupants may find evidence that they are not alone in space. On the surface of these worlds may lie the evidence of past alien visitations.

14 Communication Philosophy

SO FAR we have confined ourselves more or less exclusively to the technological aspects of interstellar communication. We have examined the fundamentals involved in searching the heavens for possible signals, and we have also explored the possibilities of finding a common medium of expression. In addition we have had a look at the electronic requirements and practicalities. Last, though by no means least, we have considered the potential of probes as a means of interstellar signalling. There is, however, another aspect that so far we have not considered at all. This, for want of a better title, we might term the philosophy of interstellar communication.

A civilization intent on establishing interstellar contacts will have allocated a specific number of high-frequency radio transmitters for this special purpose. To this end it will also have allotted a certain number of radio receivers. The alien race concerned will be able to put a figure to the number of suitable stars per unit volume of space. There will also be a fairly specific distance over which they believe they are able to transmit. As will be shown, this information is sufficient to enable the total number of suitable stars, which the particular distance involves, to be calculated.

At this point let us consider aptitude in radio communication. The distance over which a race is able to transmit should increase as its technology advances. If, for example, it is assumed that radio communication, so far as our own world is concerned, came into being around the year 1900, then it can be said that prior to this our capabilities in the field were zero and that consequently the distance over which we could transmit was also zero. As our technology in this respect progressed so also did the distance over which we were able to transmit. Eventually, if a state of perfection in radio communication is to be reached (a very theoretical premise!) the distance will have to become constant.

At this juncture another factor must be considered. Though increasing electronic aptitude makes it possible for a civilization to transmit ever deeper into space, interest in other stellar systems by this particular race will decrease in direct proportion to the increase in distance. This, of course, is due to the greater time in which the signals are in transit. Distance must therefore be restricted on these grounds. By imposing such a limitation it becomes possible to calculate on a more realistic and rational basis the approximate number of suitable star systems available as potential contact points.

If now we consider the probable number of transmitters that a civilization possesses for this purpose, we can start drawing a number of reasonably logical conclusions. Because of the very considerable expense entailed in constructing, maintaining, and programming high-power, high-frequency transmitters of the type required, it is clear that the total number on any one planet can never, relatively speaking, be really large. The number of suitable stars is, however, in comparison very large indeed. We can reason further along the same lines by assuming the number of receivers to be greater than the number of transmitters, since, of course, the former are cheaper to construct and maintain, though the total number of receivers would still be low compared with the number of appropriate stars available.

Since the ratio of stars to transmitters is of necessity very high, any race intent on establishing stellar contacts finds itself in something of a dilemma. The chance of making contacts can hardly be enhanced by the concentration of all its equipment and resources either on a solitary star or on a mere handful. To have a reasonable chance of success it should beam signals to all stars within the previously determined distance that may have life-supporting planets. We will show later, however, the almost fantastic time requirements that would be involved even if *one* day were to be devoted to each star.

It is by now perhaps worthwhile to see if all this can be expressed mathematically.

Suppose we let the number of suitable stars per unit volume be N_s^{uv} and the *total* number of suitable stars be N_s. The distance over which a race believes it can transmit will be D, a value that must be expressed in parsecs. (A parsec is equivalent to 3·26 light years and is the unit used by astronomers to express distance outside the solar system. To some extent it is displacing the better-known light year. When very considerable distances are involved the practice has some merit since, being a larger unit, it enables these distances to be indicated by less unwieldy totals. So vast is the universe, however, that even distances expressed in parsecs can be formidable.)

The total number of suitable stars can, therefore, be given as

$$N_s = \frac{4\pi}{3} N_s^{uv} D^3$$

What value, it may now be asked, can be given to N_s^{uv} (the number of suitable stars per unit volume)? The proportion of stars to be found in the immediate environs of the sun capable of supporting superior galactic communities has been estimated as 0·006 (where D is expressed in parsecs). Thus N_s^{uv} is equal to 0·006. Complete substitution in the above equation can now be effected if we are able to provide a value for D. This is important since in the equation D is raised to the third power. Any error here would be greatly magnified. (π is a mathematical constant equal to 3·14.)

The state of terrestrial radio technology at the present time could justifiably be described as lying between zero and perfection. This, it has been estimated, gives us a range (i.e., a value for D) of just over 5 parsecs (approximately 16½ light years).

For older civilizations (and therefore more technologically advanced) we can think of D in terms of 500 to 1,000 parsecs. However, distances of the order of 1,000 parsecs (3,260 light years) are unrealistic due to time lag. If we compromise to some extent by taking D as being equal to 500 parsecs (1,630 light years) and substitute in our earlier equation we have the following value for N_s (number of suitable stars within range).

Thus, $$N_s = \frac{4 \times 3\cdot14}{3} \times 0\cdot006 \times (500)^3$$

$$\therefore N_s = 3,125,000$$

that is, the total number of suitable stars within 500 parsecs (1,630 light years) is in excess of 3 million.

Suppose we put the total number of transmitters available for purposes of interstellar communication as N_t and that of receivers as N_r. We cannot, of course, say what number of transmitters (or receivers) any particular civilization will possess. Nevertheless, one point stands out clearly enough: it is unlikely ever to be very large. Such equipment, and this applies more specifically to transmitters, will be costly to construct, maintain, and operate. There is no return for such capital expenditure (at least not of an immediate kind) so that the necessary financial appropriations may be less than generous. In comparison to the number of appropriate stars, the number of transmitters must be very low indeed. It will also be lower than the number of receivers, but in this instance the disparity will be less appreciable.

Using mathematical notation then, we can say that $N_s >> N_r > N_t$ ($>>$ = much greater than; $>$ = greater than).

With D equal to 500 parsecs and N_s (number of stars) to 3 million, an arbitrary figure of 10 for the number of transmitters leads to some interesting conclusions, for example,

$$N_s = 3 \times 10^6 \text{ (3 million)}, N_t = 10$$

$$\therefore \frac{N_s}{N_t} = \frac{3 \times 10^6}{10} = 3 \times 10 \text{ (i.e., 300,000)}$$

$$\therefore N_s = N_t \times 3 \times 10^5 = 300{,}000 \, N_t$$

There are therefore, on this basis, 300,000 stars to *each* transmitter: this then is a measure of the dilemma confronting any race attempting to engage in interstellar communication. Were it to allocate to each transmitter one particular star, the chances of a successful contact would be infinitesimally low. It should direct its radio beams successively to all suitable stars within reach or at least to a high proportion of them. On this basis, however, *one day* spent on each star would entail a programme requiring 800 years. Such a period would certainly be quite unacceptable to a race like our own whose life expectancy is considerably less than this. Moreover, it is apparent that one day to each suitable star per transmitter is so inadequate as to make the exercise barely worthwhile.

Let us reassess the position on a basis of 100 transmitters to a civilization. To an advanced race of beings possessing a highly developed technology this is hardly an unreasonable figure.

In this instance then,

$$N_s = 3 \times 10^6, N_t = 10^2 \text{ (i.e., 100)}$$

$$\therefore \frac{N_s}{N_t} = \frac{3 \times 10^6}{10^2} = 3 \times 10^4 \text{ (i.e., 30,000)}$$

$$\therefore N_s = N_t \times 3 \times 10^4 = 30,000 \, N_t$$

This, then, allows for one transmitter to every 30,000 appropriate stars, or, were one day to be spent per transmitter per star, a programme that would cover 80 years. Though this is a rather more acceptable period of time, the programme itself is still virtually valueless. Much more than one day must be spent by a transmitter on each star. Suppose, however, *100 days* were to be spent directing a radio beam towards each star. This is not a long period. In fact it is still only marginally acceptable. Unfortunately the relevant programme period escalates frighteningly to over 8,000 years! On our earlier basis of a mere ten transmitters it becomes 80,000 years.

Our aim must be to rationalize, for clearly periods of these orders are useless. In two respects there must be compromise. We must restrict the number of transmitters, and we must drastically curtail the number of stars. It might be argued, in the light of the facts, that these cannot be other than retrograde steps. On the other hand, what could be more retrograde than programmes varying in length from 80 to 8,000 years that permit only brief coverage of each star?

Our policy must be to derive a programme operable within a human lifetime, a policy that covers as many stars as possible yet allots to each a period of reasonable length. Five hundred parsecs (1,630 light years) represents a compromise value for D but clearly this involves us with periods much too long by human standards. We do not know how life expectancies may vary throughout the universe but there is certain biological evidence to suggest that 50 to 200 years may be an acceptable average. If, therefore, we reduce D to 50 parsecs (163 light years) we improve the position not only from the point of view of a civilization's resources but also with regard to transit time of signal and reply.

$$N_s = \frac{4 \times \pi}{3} \times 0.006 \times D^3$$

$$= \frac{4 \times 3.14}{3} \times 0.006 \times (50)^3 = 3,125$$

$$\therefore \frac{N_s}{N_t} = \frac{3,125}{25} = 125$$

$$\therefore N_s = 125 \, N_t$$

Thus the number of suitable stars becomes 3,125, and if 25 transmitters are allocated we find that each transmitter must cover 125 stars. If 100 days is the time per star per transmitter, the overall period of the programme is 35 years.

All this constitutes the 'philosophy of the first contact'. How should we reason after the first definite link has been established? Does a race at once turn all its available transmitters on to the star from whose environs a response has been evoked, or does the positive contact lead simply to more intensive concentration on all other stars in the programme?

The latter seems the more likely course. The very fact that at last the time and distance barrier has been breached should give added impetus to the entire project, to a desire to repeat the success at other points in the heavens. A civilization achieving such a breakthrough would know beyond all doubt that it was not alone in the universe. Despite all the good grounds, cosmological, biological, and philosophical, for believing the universe to be peopled by a host of other beings, the fact cannot yet be absolutely proven. A contact, any positive contact, spells out confirmation in the boldest of symbols. A single race in the galaxy might just have been a monstrous, natural accident or an ironic cosmic joke! Confirmation of the existence of even *one* other race means that life can develop elsewhere—and if at one other point why not at many? Surely then the first contact *must* lead to intensification of the search at other points.

Initial signals sent out by an alien culture are likely to be in the nature of an electronic reconnaissance, that is, their purpose would be for others to recognize them and having done so to reply. Only then does it seem likely that information-containing signals would be sent. Were this course not adopted much time and effort might be expended on peoples either 'deaf' to the whispers from space or unable to reply. We must assume that an alien race is not likely to be interested for long in passing detailed information concerning itself to a star system that fails to respond. Were it certain its signals were being intercepted and understood this race might continue to transmit in that direction for some time. Obviously, of course, there is no way in which it could be aware of this. Another civilization fully able to respond would be of infinitely greater interest.

We have already seen how the realities of the situation must compel a race to limit rather drastically the number of stars on its 'contact' programme. Stars most likely to be given low priority would be those lying at points farthest out. A race would almost certainly direct its probing signals at suitable close stars since a shorter period would be required for response.

This brings us logically to the question of contemporary civilization distribution within our galaxy. It is virtually impossible in the light of our knowledge at the present time to arrive at a figure representing the possible average separation distance between superior galactic communities. Were it to be 5 to 10 parsecs (16–33 light years) it could be described as favourable; around 100 parsecs (326 light years) there would be obvious disadvantages.

If aliens are adopting a scanning technique in respect of close favourable stars including our sun, then their narrow radio beam may rest on each star system for only a brief spell. This could render detection and identification of such a beam a difficult matter.

Let us assume for a moment that a race of intelligent beings on Planet Epsilon Indi III directs its probing beam towards us for one day every 20 years. This is unlikely to be sufficiently long enough for us to interpret it correctly. The following day we continue to listen but hear only the sighing of the inanimate universe. The beam from Epsilon Indi III has swept on along its inexorable path and rests now upon another star system. Immediately a transmitter on Earth must get a response on its way. It does not matter if we have not interpreted the message correctly, does not indeed matter if we have doubts about its origin. What does matter is this: we *may* have intercepted a signal from an alien race. The beam – if it was one – has swept on. It may not return for a very long time, if at all. This other race will, if it receives no response, be inclined to assume that the environs of the sun are devoid of life or at least of civilizations possessing highly developed cultures and technologies. These people will certainly allow sufficient margin for their signals to reach us and for a reply from us to reach them but that may be all.

All the reasoning we have carried out so far may seem rational and logical enough to us. It is terrestrial reasoning, a product of terrestrial minds. Now, however, it becomes necessary to reflect on a possibility so far ignored or overlooked. Do communities existing on other remote and alien worlds possess differing mentalities? Are they in fact, on a different wavelength so far as mental processes are concerned? Could it be that our space-communication philosophy has an entirely wrong slant?

A race having the aptitude and ambition to seek interstellar contacts must be regarded as a highly technological one. What we have to consider is whether or not alien mental processes could lead to conclusions seemingly less logical than those reached by ourselves. Is it possible, for example, for another civilization to believe that a 120-minute beam directed towards our sun every 2·75 years represents the best method of announcing its presence – and that failure to evoke a response along similar lines proves the sun's planets to be sterile or inhabited only by morons?

Many years ago a fantastic scheme was suggested whereby the attention of the Martians was to be attracted by the construction of a gigantic diagram (either in stone or vegetation) upon a suitable portion of Earth's surface. This was to illustrate the well-known geometrical proposition of Pythagoras, known (if not loved) by every schoolboy! (The square upon the hypotenuse of a right-angled triangle is equal to the sum of the squares on the other two sides.) The meaning of this diagram, it was argued, must be as plain to alien mathematicians as to terrestrial ones. More recently, however, it has been suggested that an extra-terrestrial race might not think of representing a mathematical square by a physical one. In this case the great diagram would be quite meaningless.

Is it possible that alien mathematics, starting from different principles and under the influence of entirely different logic could have developed in a totally different way? A case might certainly be made for this sort of argument but it is difficult to see it as a very convincing one.

Suppose, for example, a stellar civilization sends out a signal consisting of four pulses as follows:

● ● ● ●

This, as yet, is meaningless. It is followed, however, an hour later by another that reads

● ● ● ● ● ●

By the time its successor comes along after another hour we are left in little doubt as to the meaning of

● ● ● ● ● ● ● ●

We can in fact now predict the form of the next signal which will be

● ● ● ● ● ● ● ● ● ●

All this is perfectly logical. One plus one gives us two *on any world*! One plus two must equal three. The 'message' is clearly intended to read

$$1 + 1 = 2$$
$$1 + 2 = 3$$
$$1 + 3 = 4$$
$$1 + 4 = 5$$

Is it inconceivable that on any planet, in any star system, in any galaxy, 1 plus 2 can equal 4 or indeed any integer other than 3?

Let us return to the relationship between a physical and a mathematical square.

PHYSICAL SQUARE   9 (MATHEMATICAL SQUARE)

On the previous page are portrayed a physical square and a mathematical one. There seems to be no relationship between the two. We know, however, that the sides of a square are by definition equal in length. Suppose we now consider the two in the form shown below.

In the physical square the length of each side has been made equal to 3 units. Three multiplied by itself, that is, 3 'squared' ($3^2 = 3 \times 3$) is, of course, 9. The mathematical square is shown as 9 units just as the physical square has been divided into 9 units or smaller squares. In other words the square root of a number when translated into dimensional physical units gives a physical square so long as the lines are at right angles to one another. This fundamental truth must surely apply on *any* planet *any*where.

Briefly, then, our case is as follows. Logic in abstract matters could conceivably vary from race to race, from world to world. Indeed on Earth this is evident, notably in the fields of philosophy, art, and so on. In matters technical and scientific we are, however, bound by certain inalienable and inflexible physical laws. These cannot and will not change and there can only be one form of logic when it comes to dealing with them. The physical material universe with which alien races must contend is the same physical universe that confronts us. Problems and conditions are the same. The solutions to these problems require that certain specific physical demands be complied with – and interstellar communication is largely a physical problem.

On our own planet we find that in art, culture, and music the peoples of the Orient and South-East Asia are distinctly different from those of Western Europe and North America. Their philosophy and temperament are in a number of respects well removed from ours. To Western ears the music of China or of India, though possessing a haunting, indefinable charm, is alien. Nevertheless an Oriental people building a jet plane or a diesel locomotive produces something that to all intents and purposes is the replica of its Occidental counterpart. The constructional problems in each case are the same and the solutions to these are dependent on the application of certain specific and quite definite physical principles. The reasoning of the Orient in such matters must of necessity be that of the West and vice versa. If it were not, then presumably the jet would not fly or the diesel locomotive move!

Admittedly comparison between the West and the Orient is a vastly different proposition from that between races light years apart, yet it seems that a similar if not identical principle must apply. We have reasoned that alien beings will sweep the heavens with their radio beams in a certain way. This reasoning, the product of terrestrial minds, is dictated by the physical laws of the universe, laws that must of necessity apply to all peoples of the cosmos. It is, therefore, at best unlikely that systematic programmes of radio sweep and search carried on by alien beings can be conducted in a manner varying greatly from that which we would employ. In many respects extraterrestrials will reason rather differently from us. When, however, they come to deal with the material universe they find themselves confronted by precisely the same difficulties and problems. These peoples may gibber rather than talk, their music may be a cacophony of weird cadences. They may play games that to us are not games, paint pictures meaningless to our eyes. But the laws of matter, of the physical universe, cannot change. In art and kindred subjects thought and reasoning may have a large measure of freedom. In scientific philosophy no such degree of freedom is permitted. The answer is either right or it is wrong. There is no alternative way, no convenient middle course.

And so, therefore, we may feel reasonably sure that alien beings having a proper awareness of the universe must, so far as the universe is concerned, reason much as we do. Some slight variations in technique and procedure are inevitable. We must at all times be prepared for practices reflecting a greater measure of racial maturity, conclusions that speak of higher technical standards, reasoning that indicates higher intellectual planes. Nevertheless there should be little in alien practice, in fundamental alien philosophy, that should really surprise us. The mental processes, the reasoning, the way of arriving at a conclusion may all differ to some extent but nothing can alter the fundamental requirements.

Thus, though the race that attempts to communicate with us is of a very different form, though its society bears little resemblance to our own, it must come to terms with the physical universe just as we must. Its searching, probing radio beams are subject to the same laws as ours.

15 The Road Ahead

IT IS now almost time to end our narrative. We cannot and certainly do not claim to have covered entirely every aspect of interstellar communication. On the contrary, we have merely scratched the surface. In doing so, however, we have come to realize how great indeed are the depths.

At present the theme of communication with peoples of the star worlds rides uneasily on the fringes of respectability. Enthusiasts should not be discouraged by this state of affairs. During the first fifty years of this century being a serious protagonist of travel to the moon or the nearer planets of the solar system was to invite ridicule, scorn, and at times even downright hostility. As late as 1940 such journeys still belonged exclusively to the realms of science fiction. Even enthusiasts felt that there they would probably remain until at least the closing years of the twentieth century. By 1950, however, there were ominous stirrings. The Second World War had come and gone, leaving in its wake a greatly enhanced and accelerated technology. On the scene had appeared both the awesome power of the atom and the tremendous potential of the rocket. Seven years later came the dawn of the Space Age. Today man has already left his footprints on the surface of the moon and is even now setting his sights on the nearer planets. Within a few decades the establishment of lunar and Martian bases is a certainty. Suddenly interplanetary travel has become respectable, even academic. As the brave new science of astronautics it has taken its place in the general framework of astronomy.

We feel certain that interstellar communication must follow a similar pattern. Thanks to the endeavours of Frank Drake and his colleagues of Project Ozma this pattern is already developing. At the same time we must not forget the less spectacular but equally worthy theoretical work carried out by a small, dedicated band of men. This is manifest in the papers and published work of Morrison, Cocconi, Bracewell, Webb, Oliver, and others. The science of interstellar communication has come to stay. It may have arrived early on the scene, before man is really ready for it. But come it has. The early years of its course may well be turbulent. It would be odd perhaps if they were not. For years it must remain not only speculative but highly controversial, yet we feel certain it will survive. Man's new and broader cosmological outlook alone should ensure this. The increasing extent of his domain among the worlds of the solar system must inevitably orient his thoughts from the affairs of one planet to those of the galaxy at large. When man finally reaches lonely Pluto he cannot for a considerable period proceed further. But on the other side of the black and terrible interstellar ocean lapping around his feet will lie other worlds, other peoples, other civilizations. He will want to hear from them. And they, we may be sure, if they are aware of his presence, will want to hear from him!

It is essential, however, that further work in this field be so defined and oriented as to preserve the aura of respectability and authenticity that has recently fallen upon it. This does not mean that it should not be popularized. The world must be made aware of its enormous potential, of its incredible fascination. What must be guarded against is the establishment of popular misconception. To the layman the idea that somewhere, somehow, a few crazy scientists are about to communicate with horrifying creatures on weird, remote worlds could all too easily be given. We must always keep in mind the host of enthralling possibilities inherent in the idea of life far out there. Our aim is to contact the star worlds, to explore fully all the possibilities – not to amplify ideas whose rightful place is between the covers of the more lurid types of science fiction! Our speculation must as far as possible be based securely on solid factual foundation.

During September 1971, an international conference was held at the Byurakan Astrophysical Observatory in Soviet Armenia. This was sponsored by the United States and Soviet Academies of Science, the Armenian Academy acting as host. The conference included historians, anthropologists, and cryptographers as well as astronomers and biologists.

In scientific circles what we have hitherto termed interstellar communication has in fact been given the rather more formal title of Communication with Extra-Terrestrial Intelligence or, since this is apparently the age of abbreviations, CETI. This seems both appropriate and significant since the star Tau Ceti has already been the subject of preliminary investigation in this respect (Project Ozma, 1960) and is, as we have seen, a likely centre of extraterrestrial and extrasolar intelligence.

During the conference the Russians reported that for some time they had been looking for extrasolar signals and had even recorded some that for a time had been considered promising. An array of radio-telescopes at four points within the Soviet Union had received a number of simultaneous pulsed signals. These unfortunately were received only during daylight hours. Since it seems highly improbable that inhabitants of a distant planet would time their transmissions to coincide with daytime in the Soviet Union it was not unreasonable to assume that these signals were terrestrial in origin and of similar nature to those dramatically received during the first hours of Project Ozma. (It would of course be most intriguing to think that remote alien beings have so effectively monitored the affairs of this planet as to be aware of when a particular region is basking in daylight!)

The final verdict of the conference was that research should be concentrated in two directions both of which were, in the opinion of the delegates, 'highly promising'. These were as follows:

(i) Searches for civilizations at a technical level comparable with our own should be put in hand.

(ii) Searches for civilizations at a technological level greatly surpassing our own should also be initiated.

It was further agreed that a wide circle of specialists, ranging from astrophysicists to historians, should participate in the planning and organization of such research. Recommendations were threefold:

1. A search for signals and for evidence of astro-engineering activities in the radiation of a few hundred chosen nearby stars and of a limited number of other selected objects covering the visible to decimetre wave range should be instituted, using the largest existing astronomical instruments.

2. A search for signals from powerful sources within galaxies of the 'local' group, including strong impulse signals, should be put in hand.

3. Exploration of the region of minimum noise in the sub-millimetre band should also be organized in order to determine its suitability for 'observing' extraterrestrial civilizations.

The conference also classed two other lines of research as 'desirable':

(a) Studies should be conducted and designs drawn up with a view to the construction of certain specific types of telescope:
   (i) A decimetre wave radio-telescope having an effective area greater than 1 km^2.
   (ii) A millimetre wave telescope with an effective area greater than 10,000 m^2.
   (iii) A sub-millimetre wave telescope having an effective area greater than 1,000 m^2.
   (iv) An infrared telescope having an effective area greater than 100 m^2.
(b) A system should be considered for keeping the entire sky under continual surveillance. This would greatly enhance the prospects of (i) and (ii).

Unfortunately at the present time most of mankind's waking thoughts seem more directed towards fending off the entirely man-made 'energy crisis' and in planning, against all the odds, for a return to even greater affluence. This crisis will no doubt provide the world's governments with an excellent excuse for doing precisely nothing towards the furtherance of extraterrestrial research. Had it not existed most would doubtless manage to dream up another. We do not need to look far for substantiation of this. There was a time not so long ago when men walked regularly on the surface of the moon. Today there are none – and for the foreseeable future there aren't going to be any. *Sic transit gloria mundi!*

In June 1971 a group at the N.A.S.A. Ames Research Center under the leadership of B. M. Oliver reported that, in their opinion, the most sensitive and appropriate equipment for CETI programmes would be a large array of radio-telescopes forming a collecting surface of several square kilometres. Unfortunately, as yet, there are apparently no plans for the construction of such an array. However, the Soviet Union is presently building a large radio-telescope at the Crimean Astrophysical Observatory. When completed this will consist of a thousand metal plates set in a ring 600 metres in diameter. Part of this instrument's work will comprise routine CETI searches.

The Ames study group recommended, as might have been expected, a routine investigation of appropriate stars in spectral classes F0 to K9, but they also stressed the desirability of directing a suitable array at the M31 Galaxy in Andromeda. In this other universe, the closest to our own, there could well be as many civilizations broadcasting as in our own. As far as we are concerned such signals would all come from the *one* direction.

The conference in Armenia agreed that the cost of such schemes, although considerable, is not prohibitive in relation to the world's economy and resources at the present time, the more so when the possible benefits of tuning in to a superior civilization are considered. So important is this matter, said the delegates, that it should be done in the name of all mankind and be the subject of international coordination. To assist realization of this worthy ideal the conference proposed the formation of an international committee to coordinate programmes and promote progress. As an interim working group the delegates proposed Frank Drake, Philip Morrison, B. M. Oliver, and Carl Sagan of the United States, N. S. Kardashev, I. S. Shklovsky, G. M. Tovmassyan, and V. S. Troitsky of the Soviet Union, with R. Pěsek of Czechoslovakia. The result of all this activity will be eagerly awaited.

Already it has been suggested that in the near future it may be possible to secure a Ph.D. degree in interstellar communication. Though this may represent an over-optimistic outlook at the present time, the realization of such an ideal may be somewhat closer than most of us imagine. The subject, as we have just seen, has already become a high-priority matter with a number of eminent, capable, and highly respected scientists.

The position is rendered unique, almost bizarre we might say, by the fact that to date there is really so little to go on. As yet we have a science without a true discipline. Nevertheless its growth could be surprisingly rapid. Interstellar communication, although it must be regarded as the child both of astronomy and of electronics, is very much more than that. It is something that could ultimately impinge on the lives and destinies of each and every one of us, on our civilization (such as it is!), our society, our children, and our future. Directly or indirectly it could link up with all facets of our existence here on Earth. Much of the knowledge that we have acquired down through the ages contributes to our speculations on the subject of extrasolar life.

Receipt of a clear and definite signal from a distant alien civilization must inevitably have a much greater impact on our society than a landing by astronauts on the moon or on Mars. This does not imply that the latter are not mighty landmarks in the history of our world and of our race. A jump to the moon is after all the first short step in man's eventual journey to the stars. It is the initial link in a chain of progress whose ultimate one it is impossible to see. We appreciate, however, that on the moon there are no intelligent beings, no other civilizations. Half a century ago the imminent prospect (relatively speaking) of a landing on Mars would have conjured up marvellous visions of a meeting between ourselves and the gifted, master race of technologists and engineers then confidently believed to inhabit the legendary red planet. Today, wiser if sadder, we realize that lichens and perhaps a very low form of insect life are about all we can expect to find confronting us on that hitherto mysterious world. Irrespective of where we go within the solar system it is unlikely that we will come upon intelligent beings, other than on our own planet. When the days of such achievement have come, man will find that he is still alone. Only interstellar travel can possibly bring him into actual physical contact with alien peoples. And this, the ultimate dream of all space travellers, is something that, because of its unique character, can be accomplished only at a remote future period in the history of our race. So man, despite his spaceships to the planets, his sophisticated space liners traversing the entire solar system, will still be alone. Interplanetary travel will not, cannot change that! The soft, faint whisper in a terrestrial antenna, the stirring of a weak, almost indiscernible current in a network of wire, could in the end prove greater than the most mighty interplanetary space vessels. On that day an alien hand will have reached out of the darkness, out of the terrible remoteness, and clasped our own.

Throughout these pages we have based most of our premises on the assumption that other races in the cosmos, once they are aware of our existence, will automatically wish to communicate with us. Must this necessarily be so? To suggest seriously that some might not may seem almost heretical. We on this planet have had an advanced technology of a sort for only a very limited period. There may well be races out among the limitless star dust who have had this for millennia. Such peoples could for long have been proud possessors of an intricate intersteller network – members of a polished and highly sophisticated galactic club. That we might be regarded as a sort of cosmic kindergarten is by no means impossible. Man may have a very vital lesson to learn – that he is neither as important nor as omnipotent as at times he seems tempted to believe.

In view of all the difficulties and technical problems that interstellar communication involves, however, it is hard to believe that the peoples of other star systems would ignore us entirely. After all, there cannot be so many cosmic races in close proximity that the few who have established a communications network will be prepared to disregard the first faltering whispers from another. Earth could well prove a convenient relay station receiving, amplifying, and re-transmitting signals between planets of widely separated systems. It seems reasonable to assume that we would be accepted, and readily – but as novices! We can expect no distinctions, no favours from our galactic kindred. We should never commit the grave error of looking for them – or feel slighted if they are not forthcoming. We are just one other galactic race – no more, no less!

In this great and exciting new era it will not be sufficient for man merely to play the role of passive onlooker. He must learn to speak as well as listen. The star peoples cannot be expected to know of us if they do not hear our voice. This too must neither be forgotten, nor in the interests of expediency disregarded. Listening, we realize, is both exciting and fascinating. It possesses a distinct mystique and, therefore, a strong appeal. Transmitting information falls into a rather different category. It is possible to engage in this for long periods without a single iota of tangible result. For all we know our signals could be flowing out endlessly into space without reaching a single comprehending ear. Once more we find ourselves confronted by the effects of sheer unimaginable distance. A reply, if there is to be one, cannot in any conceivable circumstance be immediate. It will come not in hours, or days, or months, but in *years*. Our signals might indeed flow on unheeded into the cosmic depths. During the first years it will be quite impossible to tell. We must assume that initially at least the odds will be heavily stacked against us. We will be required to remain buoyed by the thought that sooner or later our message will reach comprehending alien ears, that ultimately a hand, or paw, or claw, will reach out to transmit a reply! If, of course, alien signals have first been received by us we are in possession of a tremendous incentive. Not only will we know that another intelligent and advanced civilization exists, we will also know just where to direct our signals. Let us therefore prepare now to build the great transmitters, erect the huge antennae, prepare to make the galaxy aware of our existence. Let us study how to send – and what to send. The chances of success may at first be low, but if we do nothing they will be nought. In that event there will be no first rung on the ladder of galactic status for us. Quite simply we will have failed to earn it. And the other races of the universe will not be guilty of ignoring us – they will merely be unaware of our existence!

It must be conceded of course that unanimity of opinion regarding interstellar communication is not always in evidence. This was abundantly clear in the course of a symposium entitled 'Life Beyond Earth and the Human Mind', held at Boston University in the late autumn of 1972. In this a panel of four eminent scientists were joined by a theologian; their assignment: to consider, among other things, the question of galactic civilizations and the merits (or otherwise) of seeking to communicate with such societies.

Astronomer Carl Sagan argued strongly in favour of what he termed an 'exhilarating prospect' and expressed the hope that governments would soon appropriate the necessary funds with which to begin the long and arduous search of the heavens. Already, he contended, terrestrial society had probably announced its presence to other civilizations unwittingly by virtue of its high frequency radar and military signals.

On the other hand Nobel prize-winning biologist George Wald, of Harvard University, saw contact with a superior civilization as rather a nightmarish prospect and merely another example of the misuse of science. These doubts were shared by author and anthropologist Ashley Montagu who felt that until man had learned properly to communicate with himself it would be wiser for him to be isolated from cosmic contacts.

Physicist Philip Morrison, formerly at the Massachusetts Institute of Technology, adopted a more realistic and practical viewpoint. Stressing the immensity of interstellar distance he was quick to point out that the *current* affairs of a planet could only, at best, be of academic importance.

'I do not see how tragic circumstances today are going to affect this,' he said, 'for the messages Earth receives will probably take many years to decipher and understand.'

Krister Stendahl, Dean of the Harvard Divinity School and a leading theologian had few real fears.

'If anybody is ready it is the theologians,' he declared. 'My first reaction is that it's great – because it is great when God's world gets a little bigger and I get a bigger view!' He went on, 'The growing awareness of cosmic cohabitation is enormously important for me if it fits well into a growing knowledge of God's world. It is highly probable that we are only one possible such civilization. For that to sink in – that man is only one part of the cosmos – in his consciousness – is a great achievement.'

Stendhal maintained that the image and viewpoint projected by Montagu had its roots in fear. 'When he is afraid,' he declared, 'man is a very vicious human being – that is why increased knowledge is not only interesting but is the road to learning what to fear and what not to fear.'

Carl Sagan, in reply to the dissentients, questioned whether Earth would morally contaminate another civilization and conversely whether that civilization would exploit Earth.

'The alien civilization would have to be at least our equal to have the radio technology to reach us. So we would be the low man on the exploitation totem pole. I cannot get very worried about our destroying them.' Neither was Sagan worried about danger to Earth because 'our planet is insulated by the cosmic quarantine of the enormous distances between stars'. He continued, 'I cannot believe it just right to sit here in one corner and not explore!'

More recently still, Freeman J. Dyson, already renowned for his belief in and views on galactic life, had something further yet to add. Speaking at the Institute of Advanced Study in Princeton, New Jersey, in November 1972, he struck a surprisingly sombre note. 'I have the feeling,' he warned, 'that we are going to discover things that are not so pleasant, particularly since the activities we are likely to discover first are highly technological activities. We are more likely to discover first the species in which technology has got out of control, a technological cancer spreading through the galaxy. We should be suitably alarmed if we discover it and take our precautions. It is just as well to be warned ahead of time.'

Despite his prophetic and almost Wellsian overtones, Dyson had no doubts about the correct course of action. To the suggestion that it might therefore be better to ignore any signals from the far reaches of outer space he said: 'I would regard that as the ostrich sticking his head in sand. If these things are there it is better that we know. It is no good trying to hide!'

In all this, of course, we are continuing to assume that other races do exist, that we are not alone in the universe. There may be some, perhaps even many, who disagree with this seemingly arbitrary premise. Others again might be prepared to accept the idea more readily were the merest wisp of positive, tangible evidence to be forthcoming. It is just this very evidence that interstellar communication seeks to provide. Meanwhile it is necessary that we carry on without it. It remains our steadfast belief that there is no real reason for assuming our terrestrial civilization to be unique, that we are in no sense specially privileged. We need only look up from this little planet of ours, from this speck of inconsequential cosmic dust, and gaze into the great black star-powdered yonder. Are we alone? All reason shouts 'no!' To claim without evidence that we are, seems a million times more illogical than to claim we are not. It is impossible for any of us to say when the first contact with an extraterrestrial race will be made. All we can say is this: we believe it *will* come, for we are not alone in this galaxy of ours. Somewhere in the future the long dusty road which is our history will intersect, perhaps even merge, with that of another race domiciled among the stars. In the heavens lies the ultimate destiny of man. To attain it he requires only faith. Intelligence, ability, determination, courage – these essential attributes are his already. It is faith that even now is about to carry him to the planets, to our sister worlds. It is faith that will finally put him in touch with the many peoples of the universe. It is faith that, in God's good time, will take him to the stars themselves.

Sic itur ad astra – such is the way to immortality!

Epilogue

FOURTEEN years were to elapse before a more 'active' form of 'Ozma' was to be tried. In the interval much happened. Men walked several times upon the moon and returned unscathed. Several of the nearby planets were scanned at close range. A probe left the solar system to journey forever among the stars. 'Fair stood the wind' for exploration of the universe – or so it seemed. Came 1974 and suddenly the 'happy days' were over. Crises, economic, social and international, came starkly on the scene. Man had turned inward yet again. Once more his universe was bounded by the horizon of his own little and rather pathetic world. The new Renaissance had become the false dawn. In the circumstances it is probably not short of remarkable that 1974 was singled out for an event of some importance in the realms of cosmic adventure.

Project Ozma 1960 had been a passive programme in which for a period of a few months a listening watch was kept on two likely stars. No attempt had been made to *send* a signal. In 1974 this omission was made good, though in a manner which surprised many.

The method adopted was that already described in Chapter 11 ('Towards a Cosmic Tongue'). The prime number grid in this instance comprised 73 vertical and 23 horizontal units. The message of 1,679 parts (23×73) was transmitted over a period of 169 seconds from the giant 1,000-foot radio-telescope at Arecibo in Puerto Rico. On this occasion the frequency employed was 2,380 mc/s (a wavelength of 12·6 cms) and not 1,420 mc/s.

The 'message' transmitted was interesting, ingenious – and complex (Fig. 31). It was devised by members of the National Astronomy and Ionosphere Center of which the Arecibo installation is part. Frank Drake claims that a number of people were able to unravel the message without undue difficulty. This may be so, but to the writer it seems unduly complicated. Admittedly the information which it seeks to impart – or at least certain parts of it (for example, atomic numbers, DNA helix structure) – is not in itself particularly straightforward. Representation of the human form comes over quite clearly (despite an unhappy resemblance to 'Frankenstein' on the prowl!). Portrayal of the solar system and of the Arecibo radio-telescope are also fairly clear. Nevertheless any aliens who intercept and examine this will require to have fairly high I.Q.s!

The message was directed not towards a single star but to an entire cluster of stars (M13 in Hercules) roughly 24,000 light years distant. This cluster contains some 300,000 stars and at this range it is entirely encompassed by the Arecibo beam – or will be in 24,000 years or so! Carl Sagan gives an optimistic (though quite credible) estimate of the result. In his opinion there exists a one-in-two chance of a civilization being there to receive it (in 24,000 years!).

It is rather difficult to arrive at the logic of beaming a message of this nature to stars so incredibly remote. After all, we cannot expect a reply for 48,000 years. This does seem rather long! The spatial 'geometry' of this beam certainly allows us to embrace a vast number of stars at one fell swoop, but the 'temporal' geometry – 24,000 years – must be regarded as the inevitable cost.

The entire exercise would seem to have created a certain measure of ill-feeling within the world scientific community because of the apparent secrecy with which the project was carried out. This no doubt arises from the 1971 CETI decision that any such projects should be carried out by representatives of mankind as a whole. In defence Frank Drake maintains that this was merely a trial run and of no great practical importance but others suggest that an unfortunate precedent has been set, that a strictly unilateral decision was made at Arecibo.

The whole affair sounds suspiciously like a storm in a tea-cup. The crux of the matter would seem to be that at least the thing got done – and apparently with the minimum of fuss and argument. On paper the deliberations and conclusions of august international bodies are all very well. In practice there is all too often a great deal of talk but little commensurate action – and

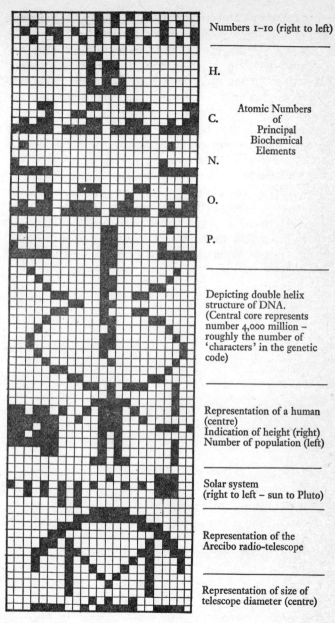

Figure 31

when the action does eventually take place it is only after a great deal of wrangling and delay.

There would appear to be a very definite case for international agreement and cooperation in respect of lunar and planetary exploration. In the realm of interstellar communication this seems considerably less important. The main point is that the thing gets done – who does it, how, where, and when is surely of much less significance. Let us have international agreement and harmony in such matters by all means – though only if this does not involve endless debate, argument, and recrimination. But – if one nation does go it alone it *is* important that details and results (if any) be regarded as the property of mankind as a whole. In other words let us cut all the red tape and get on with it. The late Winston Churchill once said, 'Never mind about the difficulties. They will speak for themselves.' This is very true. It is equally essential that we do not add quite unnecessary difficulties – there are plenty already!

Bibliography and Further Reading

ABELSON, P. H., 'Amino Acids Formed in Primitive Atmospheres', *Science*, 124 (1956), p. 935.

ALFVEN, H., *On the Origin of the Solar System*, Oxford, Clarendon Press, 1954.

ASCHER, R. AND M., 'Interstellar Communication and Human Evolution', *Nature*, 193 (1962), p. 940.

BELITSKY, B., 'Signals from Other Worlds', *Spaceflight*, January 1972, pp. 17–18.

BRACEWELL, R. N., 'Communication from Superior Galactic Communities', *Nature*, 186 (1960), p. 670.

BRIGGS, M. H., 'Detection of Planets at Interstellar Distances', *Journal of British Interplanetary Society*, 17 (1959), p. 59.

——. 'Superior Galactic Communities', *Spaceflight*, March 1961, p. 109.

CADE, C. M., *Other Worlds Than Ours*, London, Museum Press, 1966.

CALDER, N., *Radio Astronomy*. London, J. M. Dent, 1958.

CAMERON, A. G. W. (ed.), *Interstellar Communication*, New York, Benjamin Inc., 1963.

CHAPMAN, R., *Unidentified Flying Objects*, London, Arthur Barker Ltd, 1969.

COCCONI, G., AND MORRISON, P., 'Searching for Interstellar Communications', *Nature*, 184 (1959), p. 844.

DÄNIKEN, E. VON., *Chariots of the Gods*, London, Souvenir Press, 1969.

——. *Return to the Stars*, London, Souvenir Press, 1970.

DRAKE, F. D., *Intelligent Life in Space*, New York, Macmillan, 1967.

——. 'Project Ozma', *Physics Today*, 14 (1961), p. 40.

——. 'How Can We Detect Radio Transmissions from Distant Planetary Systems?' *Sky and Telescope*, 19 (1959), p. 140.

DYSON, F. J., 'Search for Artificial Stellar Sources of Infra-Red Radiation', *Science*, 131 (1960), p. 1667.

EHRICHKE, K. A., 'Astrogenic Environments', *Spaceflight*, January 1972, pp. 2–13.

FORT, C., *The Books of Charles Fort*, New York, Henry Holt & Co., 1941.

GLASBY, J. S., *Variable Stars*, London, Constable, 1968.

GURLT, C., 'Fossil Meteorite Found in Coal', *Nature*, 35 (1886), p. 36.

HALDANE, J. B. S., 'The Origins of Life', *New Biology*, 16 (1954), p. 12.

HASSELSTROM, T., AND HENRY, M. C., 'Synthesis of Amino Acids by Beta Radiation', *Science*, 125 (1957), p. 350.

HOERNER, S. VON., 'The Search for Signals from Other Civilisations', *Science*, 134 (1961), p. 1839.

——. 'The General Limits of Space Travel', *Science*, 137 (1962), p. 18.

HOGBEN, L., 'First Steps in Celestial Syntax', *Journal of British Interplanetary Society*, 11 (1952), p. 258.

HOYLE, F., *Frontiers of Astronomy*, London, Mercury Books, 1961.

LAWTON, A. T., 'The Interpretation of Signals from Space', *Spaceflight*, April 1973, p. 132.

LOWELL, P., *Mars and Its Canals*, New York, Macmillan, 1906.
——. *Mars as the Abode of Life*, New York, Macmillan, 1908.

LUNAN, D. A., 'A Space Probe from Another World', *Spaceflight*, April 1973, p. 122.

MACVEY, J. W., *Alone in the Universe?* New York, Macmillan, 1963.

——. *Journey to Alpha Centauri*, New York, Macmillan, 1965.

——. 'Creation's Dawn', *Spaceflight*, November 1966, pp. 378–412.

——. 'Interstellar Beacons', *Spaceflight*, January 1972, pp. 14–25.

MILLER, S. L.,' Production of Amino Acids under Primitive Earth Conditions', *Science*, 117 (1953), p. 528.

MOORE, P. A., AND JACKSON, F., *Life in the Universe*, London, Routledge & Kegan Paul, 1962.

OLIVER, B. M., 'Radio Search for Distant Races', *International Science and Technology*, October 1962, pp. 55–60.

OPARIN, A. I., *The Origin of Life on Earth* (3rd ed.), New York, Academic Press, 1957.

OVENDEN, M. W., 'The Origin of the Solar System', *Discovery*, 21 (1960), p. 2.

——. *Life in the Universe*, New York, Doubleday & Co., 1962.

PAWSEY, J. L., AND BRACEWELL, R. N., *Radio Astronomy*, London, Oxford University Press, 1955.

SHKLOVSKY, A., 'Is Communication Possible with Intelligent Beings on Other Planets?' *Piroda*, 7 (1960), p. 21.

SOAL, S. G., AND BATEMAN, F., *Modern Experiments in Telepathy*, London, Faber, 1954.

STRONG, JAS., *Flight to the Stars*, London, Temple Press, 1965.

STRUGHOLD, H., *The Red and the Green Planet*, London, Sidgwick & Jackson, 1954.

STRUVE, O., *Stellar Evolution*, Princeton, Princeton University Press, 1950.

SULLIVAN, W., *We Are Not Alone*, London, Hodder & Stoughton, 1965.

SU-SHU, HUANG, 'Problem of Life in the Universe and the Mode of Star Formation', *Publications of the Astronomical Society of the Pacific*, nos. 421 and 422 (1959).

——. 'Occurrence of Life in the Universe', *American Scientist*, 47, (1959), p. 397.

——. 'Life Outside the Solar System', *Scientific American*, April 1960, p. 55.

——. 'Life Supporting Regions in Vicinity of Binary Systems', *Publications of the Astronomical Society of the Pacific*, 72 (1960), p. 106.

——. 'Limiting Sizes of Habitable Planets', *Publications of the Astronomical Society of the Pacific*, 72 (1960), p. 489.

TOWNES, C. H., AND SCHWARTZ, R. N., 'Interstellar and Interplanetary Communication by Optical Masers', *Nature*, 190 (1961), p. 205.

'The Discovery of Planetary Companions of Stars', *Yale Scientific Magazine*, December 1963, pp. 6–8.

'Planetary Systems Associated with Main Sequence Stars', *Science*, 11 September 1964, pp. 1177–81.

'Are We Being Hailed from Interstellar Space?' *Fortune*, March 1961, p. 144.

TOWNES, C. H., AND SCHWARTZ, R. N., 'Relativity and Space Travel', *Proceedings of the IRE* (Institute of Radio Engineers), June 1959, pp. 1053–61.

'Direct Contact Among Galactic Civilizations by Relativistic Interstellar Space Flight', *Planetary and Space Science*, 11 (1963), pp. 485–98.

'Eye on the Future', *Nature*, 9 February 1973, p. 363.

VAN DER POL, P., AND STÖRMER, C., 'Short Wave Echoes and the Aurora Borealis', *Nature*, 122 (1928), pp. 681, 878.

WALKER, J. G., 'The Search for Signals from Extra-terrestrial Civilizations', *Nature*, 9 February 1973, pp. 379–81.

Index

Absorption lines, 205, 206, 207
Alabama, 221
Aldebaran, 59, 223
Alpha Centauri, 100, 122, 124, 154, 160, 161, 163, 164, 168, 169, 191, 193, 216, 217, 223, 228
Alpha Herculis, 60
Alpha Piscium, 60
Altair, 167
Ames Research Center, 262
Amino acids, 106, 108
Andromeda, Great Nebula in, 50, 51, 54, 136, 137, 161, 208, 263
Angular momentum, 78, 79, 80, 83-4, 87
Antares, 67, 119
Anthropocentrism, 16
Anthropomorphic, 238
Anti-matter, 123
Aquila, 167
Aquinas, Thomas, 28
Arcturus, 223
Arecibo, 271, 272
Aristarchus, 27, 30
Aristotle, 27
Armenian Academy of Sciences, 260
Arran, Isle of, 246
Aspartic acid, 108
Asratjan, E. A., 212
Associated Universities, 171
Asteroid Belt, 82, 96
Astrolatry, 25-6
Auriga, 59
Aurora Borealis, 21, 219
Aztecs, 235

Baalbeck Terraces, 237
Babylonia, 25, 232
Barnard's Star, 87, 100, 164
Bell Laboratories, 131, 202
Bergerac, Cyrano de, 33

Berkner, Lloyd, 171, 172, 173
Bermuda, 244
Beryl, 232
Beryllium, 232
Beta Centauri, 119
Betelgeuse, 58-9, 67
Bible, 230, 233
'Big Bang' Theory, 43-4
Binary star, 72, 73, 84, 87, 92, 94, 98, 168, 169
Black Holes, 227
Bondi, Hermann, 47
Boston University, 266
Boussard, 198
Bracewell, R. N., 217, 218, 219, 260
Brachiosaurus, 110
Brahe, Tycho, 31
'Bridal Veil' Nebula, 73
British Interplanetary Society, 124
Brontosaurus, 110
Brookhaven National Laboratory, 171
Bruno, Giordano, 31, 33
Byrd, Richard E., 171
Byrukan Astrophysical Observatory, 260

Callisto, 30
Canis Major, 59
Cape Kennedy, 33
Capella, 59
Carrier wave, 159
Cassiopeia, 135, 163, 191
Centaur, 137
Centaurus A, 137
Cerenkov Radiation, 226
CETI, 261, 262, 263, 272
Cetus, 22, 166
Chamberlin, Thomas C., 80
Champollion, 198
Chant, C. A., 244

279

Churchill, Winston, 274
Clarke, Arthur C., 242
Clerk-Maxwell, James, 79, 131
Coconni, Giuseppe, 155, 156, 172, 173, 174, 260
Coherent light, 202, 203
Colliding galaxies, 137, 138
Coma Berenices, 54
Comparison filter, 179
Copernicus, Nicolaus, 27, 28, 29, 30, 31, 32
Cornell University, 170
Cosmic rays, 107
Costa Rica, 235
Crab Nebula, 69, 135
Crimean Astrophysical Observatory, 263
Cryogenics, 225
Crystal lattice, 202
Curved space, 228
Cygni 6L, 87, 99, 154, 165-6, 169
Cygni SS, 70, 71
Cygnus, 58, 70, 73, 133, 135, 137, 165

Dead Sea, 234
Deneb, 58
Dicke, 47
Differentiating circuit, 179
Diffraction grating, 237
Dilation, relativistic time, 125
Diplodocus, 113
Dipole array, 142
Dish antenna, 142
Dog Star, 164-5
Doppler Shift, 155, 159, 160, 179, 211
Dorado, 57
Dragon, 175
Drake, Frank, 16, 100, 120, 170, 172, 173, 174, 175, 176, 177, 189, 192, 194, 210, 259, 263, 271, 272
'Dumbbell' Nebula, 69
Dyson, F. J., 208, 268

Eddington, Sir Arthur, 22
Edison, Thomas A., 131
Einstein, Albert, 34, 122, 126, 225, 226, 229
Electro-Acoustic Research Institute, 234
Emission lines, 205, 206

Epsilon Eridani, 101, 160, 165, 168-9, 175, 176, 182, 217
Epsilon Indi, 119, 160, 166, 168, 169, 175, 186, 187, 217
Eridanus, 165
Eureka (Nevada), 241
Europa, 30
Ewan, Harold, 170
Exhaust velocity, 123
Exponential curve, 125, 126
Ezekial, Book of, 230, 231, 232

Federal Communications Commission, 171
Feinberg, Gerald, 226
Final mass, 122, 123, 125
Flagstaff Observatory, 36
Flare stars, 136
Fontenelle, Bernard de, 33, 34
Fort, Charles, 239
Fort St Julien, 198
Frequency multiplier, 178
Fusion energy, 66

Galactic communities, 222, 253, 254
Galactic emission, 153, 158
Galaxies
  elliptical, 51, 52
  irregular, 51, 52
  spiral, 51
Galileo, 28, 29, 30, 31, 32
Gallium arsenide, 203
Gamma Andromedae, 60
Gamma radiation, 67, 107, 134, 213
Ganymede, 30
Gavreau, Vladimir, 234
Genesis, Book of, 233
Glutamic acid, 108
Godwin, Bishop, 32
Gold, 47
Gomorrah, 233, 234
Gravity/fusion balance, 66
Great Bear, 237
Green Bank (W. Virginia), 16, 100, 142, 165, 170, 171, 172, 173, 175, 177, 181, 183

Habitable zone, 97, 98, 99
Haldane, J. B. S., 105, 106, 107
Hale Telescope, 54, 139
Hals, Jørgen, 219
Harvard Divinity School, 267
Harvard University, 170, 266

Helical antenna, 143
Helium 'ash', 66
Hercules, 75
Herschel, Sir William, 34
Hertz, Heinrich, 131
Hertzian waves, 131
Hey, J. C., 133
Hiroshima, 242
Hoyle, Fred, 47
Huang, Su-Shu, 88
Hubble, Edwin, 51
Hughes Aircraft Co., 202
Huhndorff, Paul, 220
Hyper-optical, 225, 227

Implosion, 138
Incas, 236
Incoherent light, 202
Infra-red rays, 134, 208, 209
Initial mass, 122, 123, 125
Institute of Advanced Study, 268
Interferometer, radio, 143
Intergalactic space, 50
Intermediate frequency, 178–9
Interstellar gas, 49, 56, 57
Inverse square law, 128
Io, 30
Isla Verda, 241

Jansky, Karl, 131–2, 139–40, 146
Jeans, Sir James, 22, 23, 81, 82, 85
Jeffries, Harold, 22, 23
Jericho, 234, 235
Jodrell Bank, 142
Joshua, Book of, 234
Jupiter, 26, 30, 34, 41, 64, 79, 82, 83, 97, 101, 106, 135, 151, 169

Kamp, Peter van de, 87, 100
Kant, Immanuel, 23, 78, 85
Kardashev, N.S., 263
Kasantev, 233, 242
Kennelly, A. E., 131
Kepler, Johann, 28, 31, 32
KLEE, TV Station, 220, 221
KPRC, TV Station, 220, 221
Kuiper, Gerard, 87

Lake Superior, 238
Laplace, Pierre Simon de, 23, 78, 79, 85
Laser, 200, 201, 202, 203, 204, 205, 206, 207, 208, 209
Lawton, A. T., 208, 210

Lemaître, Abbé, 43, 47
Leningrad, University of, 212
Lick Observatory, 131
Lodge, Sir Oliver, 131
Lot, 233
Lowell, Percival, 36, 174
Lucian, 27, 28

M 13, 75, 272
M 31, 51, 52, 54, 136, 137, 263
M 42, 73, 136
M 87, 138
Magellanic Clouds, 54, 57
Magyar, 236
Maiman, Theodore, 202
Main sequence, 60, 61, 62, 64, 66, 67, 95
Mammals, 111
Marker frequency, 179
Mars, 15, 16, 26, 31, 35, 36, 37, 38, 39, 41, 54, 59, 64, 82, 83, 86, 89, 91, 97, 101, 151, 174, 200, 215, 246, 264
Martians, 37, 39, 195, 254
Massachusetts Institute of Technology, 203, 267
Mayas, 235
Mercury, 26, 27, 33, 34, 64, 82, 83, 97, 168
Metagalactic space, 56
Milky Way, 21, 22, 48, 49, 50, 51, 54, 56, 58, 74, 75, 92, 132, 133, 135, 136, 137, 140, 154, 174
Miller, S. L., 106
Montagu, Ashley, 267
Monte Maiz, 241
Moore, Benjamin, 105
Moore, Patrick, 91
Morrison, Philip, 155, 172, 173, 174, 260, 263, 267
Morse Code, 197
Moulton, T., 80
Mt. Hamilton Observatory, 34
Mt. Palomar Observatory, 34, 54, 87, 139, 203
Mt. Wilson Observatory, 34, 51
Mullard Radio Observatory, 142
Multiple star, 72, 73, 87, 92, 94, 98, 167

N.A.S.A., 216, 262
National Astronomy and Ionosphere Center, 271

281

National Radio Astronomy Observatory, 17, 100, 170, 171, 173, 174, 189
Nazca, 236
Neanderthal man, 229
Nebuchadnezzar, 232
Neptune, 41, 79, 83, 101, 106
Newton, Sir Isaac, 31, 34
Nice Observatory, 37
Nile, River, 198
Noctilucent clouds, 244
'Non-space', 228
Northern Cross, 165
Novae, 68, 69
Nubecula Major, 57
Nubecula Minor, 57

Oberth, Hermann, 229
Ohio State University, 143
Old Testament, 230
Oligocene Age, 229
Oliver, B. M., 260, 262, 263
Omicron Eridani, 167
Oparin, A. I., 105, 106, 107
Opiuchi, 70, 60, 87, 154, 169
Opiuchus, 136
Orion, 58, 59, 166, 224
Orion, Great Nebula in, 73, 136
Ozma, Project, 100, 120, 139, 140, 165, 169, 172, 173, 174, 177, 178, 179, 180, 181, 182, 183, 210, 211, 259, 261, 271
Ozone, 105

Paleozoic era, 105, 246
Parabolic reflector, 142, 150, 155
Parsec, 162, 167, 249, t53
P C J J, Radio Station, 219, 220, 221
Perrotin, P., 37
Pesek, R., 263
Photon drive, 227
Pillars of Hercules, 28
'Pioneer' spacecraft, 216
Planetesmal hypothesis, 80, 81, 82
Pleiades, 25, 74
Pleistocene period, 240
Plutarch, 27, 28
Pluto, 54, 77, 82, 97, 149
Pope, Alexander, 119
Population I stars, 57
Population II stars, 57
Porphyry, 241
Primordial nebula, 86, 88

Proctor, Richard A., 36
Procyon, 166
Propellant, 122
Protein, 108
Protoplanet, 86
Proxima Centauri, 58, 74, 100, 160, 164, 165
Ptolemy, 29, 30
Puerto Rico, 271
Pythagoras, 195, 254

Quartz, 109, 112, 240
Quartz-crystal oscillator, 178
Quito, 236

Radio star, 137, 144
Reber, Grote, 132
Red dwarf, 98, 164, 165, 167
Red giant, 52, 57, 64, 65
Relativity, theory of, 122
Renaissance, 27, 129, 271
Rest mass, 226
Rigel, 59
Rosetta Stone, 198
Ruby, 202
Ryle, Martin, 138

Sagan, Carl, 229, 263, 266, 272
Sagittarius, 132
Salzburg Museum, 239, 240
San Francisco Academy of Science, 240
San Jose, 235
Saturn, 26, 34, 41, 79, 82, 83, 84, 97, 101, 102, 106, 151
Schiaperelli, Giovanni, 35
Schwartz, R. N., 202
Selenites, 38
'Seven Sisters', 74
Shepherd, L. R., 124
Shklovsky, I. S., 229, 263
Siberian Meteorite, 242
Sigma Draconis, 175
Sigmoid, 126
Signal band filter, 179
Sirius, 41, 59, 61, 68, 74, 164, 165, 167, 204
Sodom, 233, 234
Southern Cross, 163
Soviet Academy of Sciences, 212
Spectral class, 60, 93
Spectrum, 204, 205, 208, 211
Sputnik I, 120

Stanford University, 217
Star cluster, 52, 56, 57, 72, 73, 74, 272
  globular, 57, 73, 74, 75
  open, 57, 73, 74
'Steady State' theory, 47
Stendahl, Krister, 267
Stormer, Carl, 219
Struve, Otto, 17, 170, 173, 174
Sugar Grove, 150, 155, 181
Super giant, 60, 61, 62, 67, 68
Super nova, 69, 135, 136
Superheterodyne, 178
Synchronous detector, 180

Tachyon, 226
Tachyon drive, 226
Tau Ceti, 101, 119, 160, 161, 166, 168, 169, 175, 176, 182, 217, 223, 261
Taurus, 69, 74, 135, 136
Technetium, 207
Telepathy, 211, 212, 213
Tertiary coal deposit, 239, 240
Tibet, 240
Tokyo, 221
Tovmassyan, G. M., 263
Townes, C. H., 202
Treasure City (Nevada), 240
Troitsky, V. S., 263
Tucana, 57
Tunguska, 242
Tyrannosaurus, 110

U Cephei, 72
U.F.O., 218, 241, 245, 246
Ultraviolet rays, 106, 134
U.S. Academy of Science, 260
U.S. Navy, 142, 170
Uranus, 34, 41, 79, 83, 101, 106, 168

Varanova, 242
Variable star, 52, 70, 71, 72
Vasiliev, L.L., 212
Vega, 223
Venus, 15, 26, 27, 35, 38, 39, 41, 64, 83, 97, 105, 151, 215
Verne, Jules, 35
Vinogradov, A. P., 107
Virgo, 138
Voltaire, 41
Vortices, 85
Vulcanism, 233
Vulpecula, 69

Wald, George, 266
Webb, 260
Weizsäcker, C. von, 23, 85
Wells, H. G., 15, 36, 37, 38, 49
White dwarfs, 52, 59, 61, 62, 65, 167
Wilkins, 32
World War II, 259

X-rays, 134, 213

Zeta Cancri, 60

COUNTER-CULTURE

AWOPBOPALOOBOP ALOPBAMBOOM Nik Cohn 50p
The original ultimate celebration of rock music: Pop from the beginning. Illustrated.

THE CENTRE OF THE CYCLONE John C. Lilly 50p
An autobiography of inner space. 'Within the province of the mind, what I believe to be true is true or becomes true within the limits to be found experientially and experimentally.' *John Lilly*.

DRUGS OF HALLUCINATION Sidney Cohen 60p
A lucid account of the discovery and first synthesis of LSD, its use and dangers in experimental psychiatry and self-induced transcendental experiences.

THE FEMALE EUNUCH Germaine Greer 50p
The book that caused a revolution, the central focus of the Women's Liberation movement.

FOLK DEVILS AND MORAL PANICS Stan Cohen 50p
Teddy Boys, Mods and Rockers, Hell's Angels, football hooligans, Skinheads, student militants, drugtakers: These are the folk devils of our time. A classic study of deviancy sociology. Illustrated.

SCIENCE

ALBERT EINSTEIN Banesch Hoffman £1.00
Written with the cooperation of Einstein's personal secretary, this is the most authoritative account of the 20th Century's greatest scientist. Illustrated.

THE ALCHEMISTS F. Sherwood Taylor £1.25
Before it became a branch of the occult, alchemy was in the forefront of the search for human knowledge and lead to the founding of modern chemistry. Illustrated.

BODY TIME Gay Gaer Luce 75p
The cosmic rhythms and cycles of the body in tune with nature and the brain.

THE CENTRE OF THE CYCLONE John C. Lilly 50p
An autobiography of inner space. 'Within the province of the mind, what I believe to be true is true or becomes true within the limits to be found experientially and experimentally.' *John Lilly*.

DICTIONARY FOR DREAMERS Tom Chetwynd 60p
A comprehensive key to the baffling language of dream symbolism. Over 500 archetypal symbols give essential clues to understanding the ingeniously disguised, life-enriching, often urgent messages to be found in dreams.

A DICTIONARY OF DRUGS
  Richard Fisher and George A. Christie 95p
From everyday aspirin and vitamin, to the powerful agents prescribed for heart disease and cancer, this is a reference guide to the gamut of drugs in today's pharmaceutical armoury.

A DICTIONARY OF SYMPTOMS Dr. Joan Gomez £1.50
Although not a full alternative to medical opinion, this is a thorough-going and authoritative guide to the interpretation of symptoms of human disease.

DRUGS OF HALLUCINATION   Sidney Cohen   60p
A lucid account of the discovery and first synthesis of LSD, its
use and dangers in experimental psychiatry, and self-induced
transcendental experiences.

EARTH'S VOYAGE THROUGH TIME   David Dinely   £1.75
A revolution has taken place in geology. The first readable account
of what we now know of earth's past, present and future life.
Illustrated.

THE END OF ATLANTIS   J. V. Luce   95p
New light on an old legend. Archaeologists, volcanologists,
seismologists show a double story – a gigantic myth and a gigantic
cataclysm. Illustrated.

THE EXTENSION OF MAN   J. D. Bernal   75p
The story of the development of physics as part of man's attempt
to control his environment and sustain his own life. Illustrated.

THE IMPERIAL ANIMAL   Lional Tiger and Robin Fox   £1.00
The authors assert that humans are fundamentally un-democratic,
adulterous, acquisitive and status-seeking. A controversial
biogrammer of human behaviour by two brilliant anthropologists.

LIFE ON MAN   Theodor Rosebury   60p
Man is literally covered with animal life. 'This brave and original
study from so many angles – clinical, literary, anthropological . . .
gave me great delight.' *Anthony Burgess*.

MAN AND ANIMAL   Heinz Friedrich   40p
Some of the most significant statements from the science of ethology –
the comparative study of animal behaviour – that enables the
de-coding of some of the secrets of human conduct.

MICROBES AND MORALS   Theodor Rosebury   £1.95
The strange social and medical history of attitudes towards
venereal disease.

THE MYTH OF MENTAL ILLNESS   Thomas S. Szasz   90p
'I submit that the traditional definition of psychiatry, which is still
in vogue, places it along side such things as alchemy and astrology
and commits it to the category of pseudo-science.' *Thomas Szasz*.
The book that rocked the psychiatric establishment.

ORGANIZED KNOWLEDGE   Leslie Sklair   75p
A sociological view of science and technology, an urgent
investigation of social responsibility in scientific endeavour.

PSYCHIATRY AND ANTI-PSYCHIATRY   David Cooper   50p
A radical social re-evaluation of the whole concept of 'madness'
and a new approach to the psychological problems of personal
relationships from one of Britain's leading radical psychiatrists.